BURLEIGH DODDS SCIENCE: INSTANT INSIGHTS

NUMBER 74

Optimising pig nutrition

burleigh dodds
SCIENCE PUBLISHING

Published by Burleigh Dodds Science Publishing Limited
82 High Street, Sawston, Cambridge CB22 3HJ, UK
www.bdspublishing.com

Burleigh Dodds Science Publishing, 1518 Walnut Street, Suite 900, Philadelphia, PA 19102-3406, USA

First published 2023 by Burleigh Dodds Science Publishing Limited
© Burleigh Dodds Science Publishing, 2023, except the following: Chapter 4 remains the copyright
of the author. All rights reserved.

British Library Cataloguing in Publication Data
A catalogue record for this book is available from the British Library

ISBN 978-1-78676-952-7 (Print)
ISBN 978-1-78676-953-4 (ePub)

DOI: 10.19103/9781786769534

Typeset by Deanta Global Publishing Services, Dublin, Ireland

Contents

Series list

Title	Series number
Sweetpotato	01
Fusarium in cereals	02
Vertical farming in horticulture	03
Nutraceuticals in fruit and vegetables	04
Climate change, insect pests and invasive species	05
Metabolic disorders in dairy cattle	06
Mastitis in dairy cattle	07
Heat stress in dairy cattle	08
African swine fever	09
Pesticide residues in agriculture	10
Fruit losses and waste	11
Improving crop nutrient use efficiency	12
Antibiotics in poultry production	13
Bone health in poultry	14
Feather-pecking in poultry	15
Environmental impact of livestock production	16
Pre- and probiotics in pig nutrition	17
Improving piglet welfare	18
Crop biofortification	19
Crop rotations	20
Cover crops	21
Plant growth-promoting rhizobacteria	22
Arbuscular mycorrhizal fungi	23
Nematode pests in agriculture	24
Drought-resistant crops	25
Advances in detecting and forecasting crop pests and diseases	26
Mycotoxin detection and control	27
Mite pests in agriculture	28
Supporting cereal production in sub-Saharan Africa	29
Lameness in dairy cattle	30
Infertility/reproductive disorders in dairy cattle	31
Alternatives to antibiotics in pig production	32
Integrated crop-livestock systems	33
Genetic modification of crops	34

Chapter 1

Advances in understanding pig nutritional requirements and metabolism

R. J. van Barneveld, R. J. E. Hewitt and D. N. D'Souza, SunPork Group, Australia

1 Introduction

Our knowledge of pig nutritional requirements and metabolism is ever evolving. Linking advances in our understanding to sustainable production of pork requires a more lateral perspective and centres on the primary drivers of sustainability.

Livestock production occupies approximately 75% of agricultural land (Foley et al., 2011), consumes 35% of the world's grain and produces 14.5% of anthropogenic greenhouse gas emissions (Gerber et al., 2013). The demand for meat and dairy products is forecast to increase 60% by 2050 (Alexandratos and Bruinsma, 2012); therefore, it is imperative in meeting this demand that pork production systems do so sustainably. It can be argued that pig production methods have always operated in a sustainable system, whereby the pig, the environment and the consumer have been considered. In the book *Meat: A Benign Extravagance*, the role of the pig has been shown to accumulate resources, otherwise discarded as waste, and to act as a hedge against oscillating availability and price of grain (Fairlie, 2010). However, as the demand for meat increases, we need to continue to advance our sustainable

http://dx.doi.org/10.19103/AS.2017.0013.05

pig production systems, which ensures outcomes that are 'good for the pig, good for the consumer and good for the environment'.

Sustainable agriculture concepts have been developed and continue to be redeveloped to address key issues over time. A three-dimensional model of agricultural sustainability was established by Douglass (1984) and is comprised of the following:

- The first dimension centres on food production, efficiency and profitability – producing sufficient quantities of food for consumers whilst providing sufficient income through the production chain.
- The second dimension examines resources (water, soil, nutrients, etc.) used by agriculture – the quantity and quality of which could impact on production and conversely the impact of production on those resources.
- The third dimension incorporates society – the regulatory environment in which agriculture and all other components of society operate and the expectations of the consumer on what the food should be, how much of it is needed and how it should be produced.

More recently, the United Nations Conference on Sustainable Development, Rio+20, called for an enhancement in food security and nutrition and a more sustainable agriculture, which has led to the formation of five principles for sustainable food and agriculture (FAO, 2014), including:

1 Improving efficiency in the use of resources is crucial to sustainable agriculture.
2 Sustainability requires direct action to conserve, protect and enhance natural resources.
3 Agriculture that fails to protect and improve rural livelihoods, equity and social well-being is unsustainable.
4 Enhanced resilience of people, communities and ecosystems is key to sustainable agriculture.
5 Sustainable food and agriculture requires responsible and effective governance mechanisms.

A key element of these principles for sustainable agriculture is improved efficiency of use of resources. In relation to advances in pig nutritional requirements and metabolism, this can be achieved through:

1 Maintaining consistent sow body condition through gestation and lactation;
2 Reducing variation in pig production systems;

3 Strategic use of metabolic modifiers;

4 Closer matching of nutrient requirements to feed composition;

5 Optimising utilisation of co-products not suitable for human consumption;

6 Optimising gut health and capacity for nutrient utilisation;

7 Enhanced understanding of nutrition and health interactions.

This chapter aims to consider these facets, focusing on advances made in nutritional requirements and metabolism and how these contribute to the sustainable production of pig meat.

2 Maintaining sow body condition through gestation and lactation

In pig production systems, improved efficiency of use of resources can be achieved by limiting nutritional resources directed towards the breeding herd whilst optimising output. Central to this is prevention of catabolism in the sow primarily through optimisation of intake during lactation.

With the above in mind, the apex key performance indicator for efficient use of resources in a pig production system is whole herd feed conversion. This accounts for all nutrients consumed by both the breeding herd and progeny per kilogram of pork produced. Minimisation of sow replacement through optimal gilt development, prevention of lameness and optimisation of reproductive efficiency will contribute significantly to reduced whole herd feed conversion efficiency (Fig. 1) and we advocate that a primary driver in the attainment of these objectives is maintenance of adequate sow body condition and minimisation in variation in this condition across the reproductive cycle.

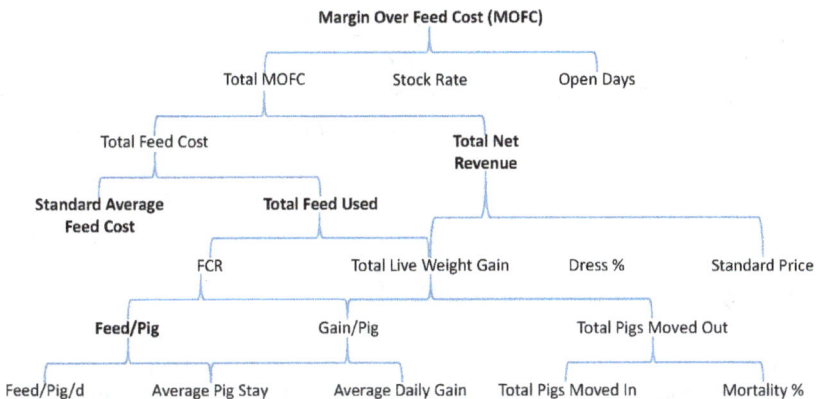

Figure 1 Apex key performance indicator for efficient use of resources in a pork production system.

2.1 Management of gilt weight on herd entry

Improper management of the sow at any stage from her selection into the herd through her subsequent parities has the potential to affect her lifetime productivity and thus her impact on the efficiency of the herd. When gilt management programmes are effective they will result in improved lifetime performance from a smaller gilt pool. Non-negotiable aspects of gilt development have been established (Beltranena et al., 2005; Williams et al., 2005), which included recognising the importance of a dedicated gilt development area, ensuring that you have enough gilts in the pool to meet requirements, breed to weight not age and on the second oestrus, and avoid slow-growing gilts. Consequently, feeding programmes during gilt development should be designed with these principles in mind.

2.2 Nutrition to prevent sow lameness

The main nutrients affecting foot health in pigs include the amino acids cysteine and methionine; the minerals calcium, zinc, copper, selenium and manganese (van Barneveld, 2010); and vitamins A, D, E and biotin (Kornegay, 1996). Biotin has been investigated for its effect on the health of pig claws with variable results being reported in response to supplementation (Brooks et al., 1977; Penny et al., 1981; Kopinski and Leibholz, 1989; Greer et al., 1991; Hamilton and Veum, 1984; Lewis et al., 1991). The availability of biotin in feedstuffs for pigs is variable with apparent ileal digestibilities of biotin from diets containing wheat and sorghum being zero (Kopinski et al., 1989).

Skeletal health is also vital to reduce lameness. Osteochondrosis has been suggested as a major cause of leg weakness in growing swine (Nakano et al., 1987) and has been linked to the premature culling of breeder stock (Gresham, 2003). The primary development of osteochondrosis in pigs occurs at a very young age; however, clinical signs of lameness are not observed until much later, up to 18 months of age (Ytrehus et al., 2007), with 80% of pigs marketed estimated to have slight to mild forms of osteochondrotic lesions (Carlson et al., 1988; Ytrehus et al., 2004). Vitamin D plays an integral role in mineral nutrition and is responsible for the intestinal absorption of minerals. Direct dietary supplementation with 25-hydroxycholecalciferol, the activated form of vitamin D that normally occurs in the liver of the pig, has been shown to promote normal endochondral ossification and inhibit the progression of osteochondrosis (Sugiyama et al., 2013).

High-producing sow lines of today are likely to have greater nutrient requirements than those sows used in past studies investigating requirements (Mahan and Newton, 1995), including those of biotin. Change in sow housing from individual stalls to group housing may result in extra attention needing to be paid to nutrient requirements to optimise sow foot health. It should be recognised that the requirement level to maintain optimum foot health may be higher than those required for optimum growth or reproductive performance. For example, Kopinski and Leibholz (1989) suggested that there is no requirement of supplemental biotin for growth of pigs; however, supplementation of 0.050-0.100 mg/kg diet is required for prevention of claw lesions. Future studies investigating the requirements of nutrients for optimum foot health should be standardised with respect to duration of the trials, floor and housing system used, and basal diet. Capacity to develop nutritional strategies to improve sow foot health will require a better understanding of the contributions of individual nutrients, alone or in combination, to foot health.

2.3 Minimising variation in nutrient intake during gestation

Nutrition during gestation is focused on meeting the requirements of the sow for maintenance and the development of maternal tissues and the conceptus (National Research Council, 2012). A higher level of nutrition in early gestation has no impact on embryo survival, but, did however, result in increase in weight and back fat (Hoving et al., 2012), whilst during late gestation, a higher feeding level also resulted in weight gain, without influencing the birthweight of piglets (Knauer, 2016). The supply of nutrients above requirements during gestation benefits only the sow and not the developing piglets, and further results in a reduction in voluntary feed intake during lactation (Weldon et al., 1994).

2.4 Maintaining body weight during lactation

A reduction in feed intake during lactation is likely to result in the mobilisation of body reserves (Bergsma et al., 2009), as the sow, in contrast to gestation, devotes all her efforts to ensuring the growth of the piglets. Body condition on exit from gestation can have an impact on this. Lean sows tend to have higher protein reserves and a higher voluntary feed intake compared with fat sows. This can be explained by high levels of exogenous protein that may actually stimulate milk production and therefore voluntary feed intake (Revell et al., 1998a). Conversely, fat sows have lower protein reserves, and thus less ability to mobilise them for milk output when supply is limited, especially in early lactation (Revell et al., 1998b). One strategy that has been employed to prevent impaired feed intake in lactation was to restrict feed intake in the immediate period both prior and post parturition (Koketsu et al., 1996); however, more recent research would suggest that *ad libitum* feeding in this period can lower mobilisation of body reserves and increase piglet performance (Cools et al., 2014).

Whilst body tissue catabolism is obviously an immediate issue for the sow and her suckling piglets, the impact on subsequent reproductive performance is considerable. Clowes et al. (2003) investigated the impact of increasing losses of protein on reproductive performance of first parity sows. Ovarian function was suppressed in sows with the greatest protein loss; they had fewer medium-sized follicles, and those follicles contained less follicular fluid, with lower estradiol and IGF-1 concentrations, translating to a lower number of piglets in the subsequent litter (De Bettio et al., 2016).

In summary, the key to a long reproductive life is to optimise the entry of the gilt into the breeding herd through meeting key weight targets, maintaining an adequate level of nutrition throughout gestation to support maintenance of body tissues and the growth of maternal and foetal tissues, and ensuring a high level of feed intake during lactation to reduce losses of body reserves and maintain good ovarian function.

3 Reducing variation in pig production systems

Variation is an often overlooked contributor to inefficient use of resources, and the nutritional interventions that can be applied to reduce variation are poorly understood or limited.

Variation is inherent in any biological system and is a challenge to manage in modern pork production businesses. In the case of the growing herd, inherent variation within a population of pigs represents a significant cost, as a result of the need to select on farm to meet market specifications, poor matching of diet specifications to nutrient requirements, grading losses, higher pre-weaning mortality and challenges associated with health management. As a consequence, any management practice that can be applied to reduce variation at the point of sale has the potential to improve the profitability and overall efficiency of a pig enterprise.

There are many factors that can influence birthweight and variability in birthweight; hence, manipulation by nutritional means alone is unlikely to resolve this. It is, however, relevant that most nutrient specifications used in diet formulations today have been derived from far less prolific sows; hence, it is possible that both the sow and foetuses are compromised, given both have an increase in requirements as gestation progresses (Ball et al., 2008). Lawlor et al. (2007) reported that birthweight, weaning weight and within-litter variation were unaffected when five different dietary digestible energy (DE) levels were provided during different gestation phases. In contrast, Wu et al. (2010) demonstrated that supplementation of gestating sow diets with specific amino acids, free arginine (to bring the lysine to arginine ratio to 2.64) and glutamine, significantly influenced litter birthweight variation. These amino acids were targeted as arginine is extensively catabolised by arginase in the small intestine, and only 60% of dietary arginine enters the portal circulation of gestating gilts (Wu et al., 2007), whilst the uptake of glutamine in the uterus of gestating gilts is the greatest among all amino acids (Wu et al., 1999) and it is abundant in placental and foetal fluids. When Wu et al. (2010) added 0.6% glutamine and 0.4% arginine to corn-soybean meal-based diets, glutamine concentrations in gilt plasma did not decline, but a marked reduction in ammonia and urea in maternal plasma and a decline of variation in birthweights among all piglets born and pigs born alive were observed. Foxcroft (2007) suggested that the smallest pigs within the litter respond the greatest to maternal nutrition interventions, resulting in a reduction in variation in birthweight as evidenced by the significant impact of exogenous somatotropin during early gestation on small pigs (Rehfeldt et al., 2001).

In addition to nutritional interventions that influence variation in birthweight *per se*, it is possible that a considerable amount of the variation in growth performance after birth may be largely determined, and essentially pre-programmed, during foetal development in the uterus (Foxcroft and Town, 2004). Foxcroft (2007) reported that the early period of myogenesis, involving the differentiation of primary muscle fibres, is generally considered resistant to nutritional manipulation, whereas the nutritional effects on differentiation and hyperplasia of secondary fibres have been demonstrated between Day 25 and Day 90 of gestation.

As discussed previously, while the greatest effects of lactation feeding will be on postnatal piglet performance and variation, nutrition of the sow during lactation can also influence variation in litter birthweights. Increased catabolism during the last week of lactation in primiparous sows is known to reduce embryonic survival and development up to Day 30 of gestation in the subsequent litter (Foxcroft, 1997). Quesnel et al. (2008) found significant relationships between sow's body condition and the coefficient of variation (CV) of birthweight. Litter heterogeneity increased with bodyweight at the beginning and end of gestation. The CV for birthweight was not linked with bodyweight gain during gestation, but with back fat gain, which was likely a reflection of changes in feed allowance during gestation as a consequence of body reserve mobilisation during the previous lactation (Quesnel et al., 2008).

Regardless of the variation in birthweight of any litter, all pig production systems should focus on optimising sow milk production and piglet colostrum and milk intake to maximise weaning weights for age. Targets of 75-85 kg of litter weight at 21 days of lactation should be a key focus for production staff, with this target representing the best insight into the effectiveness of any lactation feeding programme.

Douglas et al. (2014) demonstrated that when low-birthweight piglets are grouped and offered supplementary milk, they drink significantly ($P < 0.001$) more than those of mixed group piglets. Further, Quiniou et al. (2002) reported that small birthweights translated into small weaning weights, but this was not connected with reduced growth potential before weaning. Donovan and Dritz

(2000) removed larger pigs from the dam for a short period within 24 h after birth to allow smaller pigs in the litter uninhibited access to the dam as a means of reducing variation in weight at weaning. While no differences in average daily gain or weaning weight were detected, there were significant linear improvements in s.d. and CV of the average daily gain, but only for litters with nine or more piglets born alive. Given that smaller piglets respond well while suckling, provision of supplementary milk during the lactation period should increase overall piglet viability and weaning weight and reduce s.d. and CV. Wolter et al. (2002) offered supplemental milk replacer from Day 3 of lactation to weaning at 21 days of age. Supplemental milk significantly ($P < 0.001$) enhanced weaning weights (6.60 kg vs 5.69 kg) and also reduced the CV of weaning weights (6.62% vs 7.04%), although this effect was not significant ($P > 0.05$). Recently, Douglas et al. (2014) reported the converse. These authors were unable to demonstrate any benefit of supplementary milk on the performance of lightweight piglets before or after weaning, but it did affect their drinking behaviour and, as a consequence, the variation in bodyweight of lightweight piglets in mixed litters to slaughter was reduced.

Even when those factors in gestation and pre-weaning that we have already shown to influence variation are addressed, there is still likely to be a level of variation in a population of piglets at weaning, and variation from batch to batch. With the s.d. of the population increasing with age (Patience et al., 2004), we must also focus on efforts to reduce this immediately post-weaning.

A recent Australian study (Hewitt et al., 2015) investigated the ability to influence the performance of lightweight weaners through nutritional intervention in this period of efficient growth. It compared lightweight weaned pigs (3.8 kg at 19 days of age) with heavier pigs (5.2 kg at same weaning age) and a control group that was reflective of the normal population of weaners (4.4 kg). The lightweight pigs remained on the starter diet (15.0 MJ DE/kg, 0.8 g standardised ileal digestible (SID)/MJ DE) until they reached 30 kg liveweight, with the aim of lifting their performance to that of the heavier-weight group. Despite this significant intervention, lightweight pigs ate less and grew slower, taking an extra 13.5 days to reach market weight. Feed conversion did not differ among treatments, suggesting that boosting feed intake should lead to improvements in growth rates. Treating light- or heavier-weight pigs as a distinct group also resulted in an increase in the CV over the growth period, when compared with the control population. This has been reflected in other studies (O'Connell et al., 2005) where uniform groups at weaning became more varied over life time, while the control population became less varied, possibly due to the natural variation in growth response, differing responses to weaning and increased levels of aggression even in groups where there is no obvious hierarchy.

4 Strategic use of metabolic modifiers

The impact of technologies such as metabolic modifiers on pig production is well documented (Dunshea et al., 2005). These technologies that increase feed efficiency and lean tissue deposition while decreasing fat deposition have been developed in an effort to improve the efficiency of animal production. Some of these technologies: somatotropin, immunisation against gonadotropin-releasing factor and orally active dietary additives such as ractopamine, chromium and betaine are discussed in this section.

Perhaps the most researched metabolic modifier for improved efficiency of growth and lactation is somatotropin. The impacts of porcine somatotropin (pST) have been known for more than 70 years (Giles, 1942). Injecting growing pigs with pituitary tissue extracts containing pST increases protein deposition and decreases fat accretion (Turman and Andrews, 1955). Advances in biotechnology allowed production of pST on an industrial scale, and commercial use of bovine somatotropin and pST began in 1994 and 1996 in the United States and Australia, respectively (Bauman and Dunshea, 2010). Exogenous pST treatment has been shown to consistently improve average daily gain, feed conversion efficiency and protein deposition, and reduce fat deposition in pigs (Chung et al., 1985; Etherton et al., 1987; Campbell et al., 1988, 1989; King et al., 2000). As a technology, the growth performance of pST-treated pigs is significant, providing us in many ways a glimpse of what we might expect of pigs 10–15 generations into the future.

As discussed in the previous section, inherent variation in the growing and finishing herd represents a significant cost. Interestingly, one of the few strategies to successfully reduce the CV in finisher pig was the exogenous administration of somatotropin. Dunshea (2005) showed that pigs treated with pST had reduced CV around average daily gain compared with untreated pigs (22.1% vs 19.9% for the control and pST regimes, respectively, $P < 0.001$). The CV around P2 backfat thickness was also reduced by pST administration (16.2% vs 14.2%, $P < 0.001$).

Ractopamine, a β-agonist, is approved for use in many countries as an in-feed ingredient to increase lean tissue growth and improve production efficiency in pigs (Dunshea et al., 2005). Treatment of finishing pigs with ractopamine is well studied and has generally yielded dose-dependent improvements in daily gain, feed efficiency and carcass lean content, along with advantages in increased carcass weight, loin eye area, dressing percentage, carcass cut yields and decreased fat (Dunshea, 1993; Dunshea et al., 2005; Woods et al., 2011). These effects can be seen in all sexes, and unlike for pST, feed intake is typically unchanged or decreased only slightly during β-agonist treatment (Dunshea et al., 2016). A meta-analyses of the effect of ractopamine on carcass weight of pigs found an increase of 1.8 (n = 8 studies) and 2.4 kg (n = 19 studies) with 5 and 10 ppm ractopamine (Apple et al., 2007), which captures the range of doses used commercially. Further modelling suggests that the use of ractopamine in pig diets could have beneficial environmental impacts due to improved efficiency and a reduction in animal numbers of between 5.3 and 6.3% (for 5 or 10 ppm inclusion rates) without an impact on total meat production (Woods et al., 2011). Consequently, it is estimated that between 2.8 and 3.4 billion fewer kilograms (equivalent to 0.29 or 0.35 million ha) of corn and 0.16 or 0.34 billion fewer kilograms (equivalent to 0.059 or 0.127 million ha) of soybeans are required for US pork production each year, respectively (Dunshea et al., 2016).

Castration of entire population of male pigs, a widely practised procedure to eliminate male pheromones, commonly referred to as boar taint, and to reduce the incidence of aggressive behaviours often observed in entire male pigs, results in significant reductions in feed efficiency and excess deposition of fat. An alternative method of inhibiting sexual development and aggressive behaviours in the late finisher phase is immunisation against gonadotropin-releasing factor, referred to as immunocastration (Bonneau and Enright, 1995; Dunshea et al., 2001). This results in a reduction in plasma gonadotropins and testosterone and boar taint compounds in swine (Dunshea et al., 2001). Cronin et al. (2003) reported a reduction in both aggressive and sexual activities in immunised boars who exhibited similar activities as barrows. As a consequence, the immunised pigs spent more time eating, and increased their feed intake. A meta-analysis of 19 studies showed that, compared with non-immunised boars, immunisation against gonadotropin-releasing factor increases feed intake and average daily gain with only small increases in feed conversion rate over the finisher phase. Increases in final carcass weight and back fat depth were also shown when immunised boars were compared with entire males (Dunshea et al., 2013). Sufficient data comparing the growth performance and carcass characteristics of immunised boars with those of barrows is now available to conduct comparative meta-analyses. These analyses of up to 21 studies show that during the finisher phase, immunisation against gonadotropin-releasing factor increases feed intake and average daily gain and decreases feed conversion rate compared with contemporary barrows (Dunshea et al., 2013). From an environmental point of view, the effect of immunocastration on greenhouse gas emissions was reduced by 3.7% compared with a physical castration production system (De Moraes et al., 2013).

Further improvements in growth performance of immunocastrated pigs were also seen when used with other technologies. The growth-promoting effects of pST were enhanced when used in conjunction with immunocastration (McCauley et al., 2003; Oliver et al., 2003). Similarly, there are additive effects of ractopamine and immunocastration in pigs (Moore et al., 2009; Rikard-Bell et al., 2009).

Chromium is an essential mineral element (Mertz, 1969) with trivalent chromium (Cr^{3+}) being the most stable form occurring in nature. However, Cr^{3+}

is normally poorly absorbed and utilised with the digestibility of inorganic and organic forms of chromium being 0.5-2% and 10-25%, respectively (Mertz, 1969). It has been suggested that dietary chromium may increase insulin sensitivity in pigs (Steele et al., 1977). A meta-analysis of 31 studies in pigs showed that dietary chromium increased daily gain, feed efficiency, and carcass lean and loin muscle area while decreasing back fat thickness (Sales and Jancik, 2011). The absorption of chromium is poor, with only 0.5-3% of inorganic chromium being absorbed from the gastrointestinal tract (Dowling et al., 1989; Ducros, 1992), resulting in concentration of chromium in the effluent. Although organic forms of chromium are considered more bioavailable than the inorganic forms, only 10-25% of organic chromium is absorbed (Underwood, 1977). An alternative is nano-chromium, which has been shown to increase performance of pigs (Hung et al., 2015), but the use of nano-chromium has become a contentious issue relating to nanoparticles in the food industry (Dunshea et al., 2016).

The use of betaine in pig diets has been shown to improve growth performance by reducing the maintenance energy requirement of the animal, perhaps by reducing the need for ion pumping involved with maintaining intracellular osmolarity (Schrama et al., 2003). In addition, dietary betaine has been reported to increase protein deposition and carcass leanness and to decrease back fat (Suster et al., 2004). There is evidence that betaine has a more pronounced effect when dietary energy is limiting (Suster et al., 2004), and so it offers a means of improving growth performance through ensuring the provision of additional energy. A comprehensive meta-analyses of 15 studies on the effects of dietary betaine in pigs indicates that overall dietary betaine increased feed efficiency and carcass dressing percentage and decreased back fat thickness with no effects on daily gain (Sales, 2011).

The technologies listed in this section are just as, or perhaps more, relevant from a herd efficiency and resource use point of view today as they were when they were first utilised in pig production; however, many have reached their end from a usability point of view. Take pST for example; as a technology to improve resource use efficiency in pigs, it remains the gold standard. Yet pST use today is non-existent as a consequence of consumer and retailer pushback, as well as the need for a better delivery system that utilises less labour and is not reliant on an injection. The search for an orally administered pST delivery system remains an elusive goal.

The consumer perception of 'factory farming' has seen many of the other technologies lumped together as technologies that have been deemed to 'mess with my food' by consumer groups or retailers, with little or no supporting evidence. The resultant effect on pig production has been both the loss of key technologies that reduce environmental impact and an increase in the cost of production (Dunshea et al., 2016).

5 Matching nutrient requirements to diet specifications

Capacity to routinely, accurately and cost-effectively measure variation in the nutritional quality of feed ingredients prior to diet formulation represents a fundamental pillar of sustainable pig production worldwide, as it is the most effective means of matching nutrient requirements to diet specifications. Central to this is the capacity to rapidly measure nutritional quality of feed ingredients and up-to-date definition of nutrient requirements.

5.1 Rapid measurement of nutritional quality

Accepting the premise that measurement of 'available' nutrients will provide the greatest capacity to define the comparative nutritional value of ingredients, there is a need to prioritise which available nutrients are measured in which ingredients. This is due to the considerable cost and effort associated with establishing these analyses and the fact that it is unlikely a single analysis will be suitable for all ingredients. Apart from proximate analysis of all ingredients using either chemical or rapid techniques such as near-infrared spectrophotometry (NIRS), available energy in cereals and available lysine in oilseed meals are the highest priority for analysis of available nutrients, based on their relative proportions in pig diets and their associated variation (van Barneveld et al., 2017).

Significant debate has ensued over the suitability of DE and metabolisable energy (ME) versus net energy (NE) systems as the most appropriate measures of available energy in feed ingredients, particularly cereal grains. The latter has been touted as a better alternative on the basis that DE and ME are poor predictors of animal performance (de Lange and Birkett, 2005). This claim is unsupported by recent research (Cadogan, 2009). The digestibility of a feed is affected little by animal factors, except for young piglets and sows (Noblet and Shi, 1993; Nitrayova et al., 2006). However, NE values for a feed depend on characteristics of the feed, nutrients absorbed and the metabolic status of the animal, with different NE values derived for the same feed when consumed by animals of different genotypes, stages of growth, physiological states and climatic conditions (Kil et al., 2013). The debate about superiority of NE is irrelevant because it is impossible to measure NE at the point of feed manufacture. Furthermore, NE calculations depend on information on the DE content of a feed. Consider a NE equation reported by Noblet (2006):

$$NE \text{ (MJ/kg, DM)} = 0.703 \times DE \text{ (MJ/kg, DM)} - 0.0041 \times CP + 0.0066 \times EE - 0.0041 \times CF + 0.0020 \times ST$$

where CP is crude protein, EE is ether extract, CF is crude fibre and ST is starch (all g/kg DM). The equation shows that more than 70% of the NE contribution from an ingredient is derived from DE. Rapid analysis of DE combined with additional chemical components does allow use of NE as an estimate of available energy for an animal, but without the DE component the NE calculation would be impossible.

Further justification for the development of rapid measures of DE in cereals for pigs was provided by van Barneveld (1999). Differences of up to 3.7 MJ/kg dry matter (DM) in DE content were observed following a review of data for more than 70 cultivars of wheat. Similarly, analysis of data for more than 125 cultivars of barley revealed a range in DE estimates from 11.7 to 16.0 MJ/kg DM. The cultivar itself was shown to have a minimal effect on the availability of energy in cereals, while the cultivation conditions and agronomic practices (e.g. fertiliser rate) were reported to have a greater influence than the growing region or the growing year. Many physical and chemical factors were shown to influence DE in cereals, including protein source and type, starch characteristics, fat source and type, non-starch polysaccharide components and anti-nutritional factors. Van Barneveld (1999) concluded that the availability of energy to the animal will depend on the particular combination of these components in a grain and how they behave in the presence of nutrients from other feed ingredients. Hence, a rapid measurement of DE is best based on data derived from *in vivo* measurements.

In feed ingredients such as soybean meal and canola meal that have undergone processing, heat damage can result in an overestimation of nutritional quality due to reactions between the ε-amino group of lysine with other compounds such as carbohydrates (van Barneveld et al., 1999a). It appears that most processors are aware of the potential for heat damage during oil extraction, but quantification of these effects occurs rarely. Rutherfurd et al. (1997) developed the digestible reactive lysine assay as a means of assessing heat damage in feed ingredients. Van Barneveld et al. (1999b) applied this technique to assess the degree of heat damage in cold-pressed, expeller - extracted and solvent-extracted canola meal. Despite similar total lysine concentrations in the canola meals, true ileal reactive lysine digestibility was significantly higher in the cold-pressed meal. Reactive lysine content was shown to be a good indicator of true ileal digestible reactive lysine content (and thus the degree of heat damage). Given this relationship, application of rapid assays for total lysine and reactive lysine would allow more accurate assessment of the nutritional quality of ingredients such as soybean meal and canola meal, and account for this at the time of diet formulation and manufacture.

The need for rapid analysis of reactive lysine in oilseed meals as a routine assessment was demonstrated by Kim and Mullan (2012). A total of 209 samples collected from the major soybean meal producing countries revealed a 27% variation in reactive lysine content (21.9 g–30.1 g/kg as is basis) across all samples and 13% variation within shipment samples. This finding suggests that there is wide variation in reactive lysine content in soybean meal samples, and the use of spot sampling to analyse reactive lysine content can lead to large inaccuracies. Data analysis showed the proportion of reactive lysine relative to total lysine was not different ($P > 0.05$) due to country of origin, while reactive lysine content was significantly different between countries of origin ($P < 0.001$). Kim and Mullan (2012) concluded that the variation in reactive lysine content caused by heat processing is similar between countries but that the total amount of lysine and reducing sugars present in the raw soybeans prior to heat processing are a major source of variation. Nevertheless, the linear relationship between total and reactive lysine was weak (Slope: 6.00, intercept: 0.7152, RSD: 0.913, R2 = 0.52) and cannot be used for robust prediction of reactive lysine from analysed total lysine content.

van Barneveld et al. (1998) developed the first NIRS calibration for the measurement of DE in cereals for growing pigs, using more than 150 samples from across the world that had been subjected to *in vivo* DE analysis (Table 1). Innovative calibration development techniques were employed so that samples from various laboratories could be meaningfully included in a single calibration. This research demonstrated that DE in whole grain cereals can be measured by NIRS to an accuracy of at least ±0.38 MJ/kg, which at the time was comparable to the accuracy obtained when using *in vivo* reference methods. No significant differences in accuracy were observed between milled and whole grain calibrations or between local and global calibrations.

Further to the development of DE calibrations for cereals in pigs, van Barneveld and Ru (2003) also demonstrated that NIRS could be successfully employed to measure the total and reactive lysine content of canola meal, with obvious potential to extend this to soybean meal.

Significant investment has ensued in Australia in the past 18 years to build upon these base calibrations and we now have a practical means of rapid, online assessment of DE content in cereals used in pig diets, together with the

Table 1 Summary of calibration statistics for digestible energy (MJ/kg) in whole and milled cereal grain samples for pigs (from van Barneveld et al., 1998)

Sample set	SECV (MJ/kg)	RER	SECV/STD
Whole grain	0.518	8.8	0.49
Whole grain (IV)	0.377	12.1	0.36
Milled grain	0.596	7.7	0.53
Milled grain (IV)	0.428	10.7	0.38

SECV, standard error of cross validation; RER, range error ratio; STD, standard deviation; IV, indicator variables.

Table 2 Summary of calibration statistics for AusScan NIRS calibrations for ileal and faecal digestible energy (MJ/kg, as received in cereals, excluding oats and maize) for pigs, and total and reactive lysine in canola meal (g/kg, as received, Black et al., 2013)

Measurement	N	Mean	SD	SECV	1-VR	RPD
Ileal DE (MJ/kg)	236	11.5	1.03	0.48	0.80	2.17
Faecal DE (MJ/kg)	288	13.7	0.69	0.26	0.86	2.65
Total lysine (g/kg)	113	20.19	1.37	0.65	0.78	2.11
Reactive lysine (g/kg)	113	17.97	1.75	0.92	0.73	1.91

DE, digestible energy; N, number of samples used in the calibration; SD, standard deviation of values in the population; SECV, standard error of cross validation; VR, variance ratio; RPD, ratio of prediction to deviation (SD/SECV).

option for routine assessment of batches of heat-treated oilseed meals at the point of receival (Black et al. 2003; Table 2).

Recent advances in the development of NIRS calibrations for measurement of digestible nutrients in ingredients for pigs together with more robust NIRS equipment for online applications in feed mills have led to the collection of data that should concern all commercial nutritionists and pig producers. Online measurement of the DE and crude protein contents of wheat in a large Australian feed mill demonstrates why real-time measurement and the capacity to account for variation should exceed all other priorities in the application of feed evaluation systems (Fig. 2). Over the course of one hour of production,

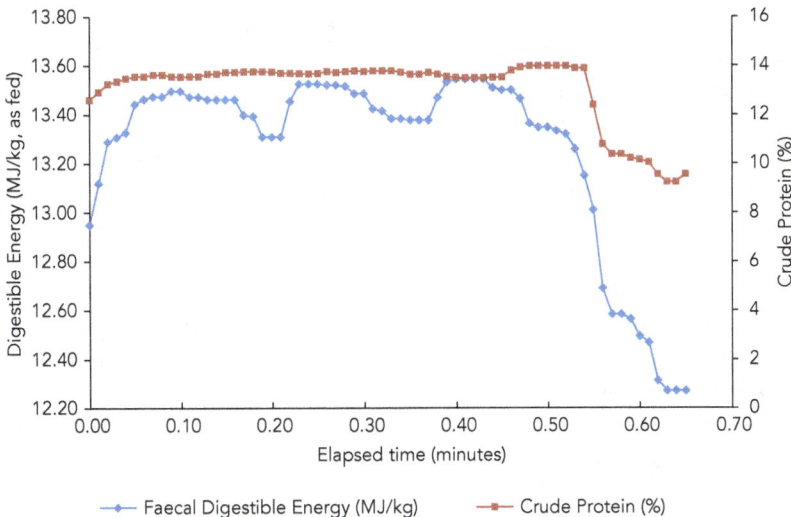

Figure 2 Real-time NIRS analysis of faecal digestible energy (MJ/kg, as fed) and crude protein (%) contents of wheat included in pig diets (D. Pearson, Feedworks Pty Ltd, pers. comm.).

the crude protein content of the wheat varied by 4% units and the DE content by 1.2 MJ/kg as received. Further, the mean DE content was approximately 0.6 MJ/kg lower than the reported book values for wheat (which would be commonly used in the diet formulation). In this particular mill, production rates are approximately 20 tonnes per hour, and it is common for large batches of single diets to be produced over this period. The online data suggests that the first 3 tonnes of feed utilised wheat with a DE content between 12.9 and 13.4 MJ/kg, the next 12 tonnes was fairly consistent at 13.4 MJ/kg, but the last 5–6 tonnes varied between 13.2 and 12.2 MJ/kg as fed. If diets were formulated as a starter feed, some piglets would receive a diet equating to 14.0 MJ DE/ kg while others would receive a diet equivalent to 14.7 MJ DE/kg, and the nutritionist would be expecting delivery of feed equating to 15.0 MJ DE/kg to these pigs. This difference in energy intake would suppress production efficiency and represent a major source of variation in pig growth which would increase as the pigs grow.

5.2 Updated information on nutritional requirements

The 11th revised edition of *Nutrient Requirements of Swine* from the National Research Council (NRC) was released in 2012. As in the 10th edition the main emphasis of this publication was to present estimates for nutrient requirements of pigs, and on available nutrient contents in feed ingredients. In addition, information was included on the pigs' need for water, co-products from the corn and soybean industries, estimation of nutrient digestibility, non-nutritive feed additives, feed contaminants, feed processing, means to minimise nutrient excretion and research needs. New chapters in this latest edition also included information regarding the nutritional importance of lipids and carbohydrates.

It has been quite some time since the nutritional requirements of the pig were sourced from actual experimental studies to elucidate the nutritional requirement of the growing pig, lactating sow and gestating sows. Instead, mathematical models to estimate nutrient requirements are commonly used (National Research Council, 2012). As summarised by de Lange (2013), models are now used to dynamically represent the partitioning of energy intake and to then estimate amino acid, phosphorus and calcium requirements of swine in a relatively disease- and stress-free environment. Similarly, for the interpretation of nutrient requirements studies, dietary nutrient levels were recalculated and, when needed, estimates of nutrient requirements were established based on a reinterpretation of the data sets. Empirical estimates of requirements for vitamins, fatty acids and minerals other than calcium and phosphorus have been updated from the National Research Council (1998; de Lange, 2013).

6 Optimising utilisation of co-products

In modern intensive production systems focused on providing people with proteins, the animal can be used to 'value-add' raw materials that would otherwise be difficult to include in a human diet. The pig excels in this context and can make use of co-products and grains that have been downgraded from human grain markets. Characterising the nutritional quality of these raw materials often demonstrates their nutritional value exceeds the physical characteristics that may have compromised their suitability as a human food product, making them comparatively valuable components of pig diets.

The FAO estimates that each year, approximately one-third of all food produced for human consumption in the world is lost or wasted (FAO, 2013). This food wastage represents a missed opportunity not only to improve global food security, but also to mitigate environmental impacts and resources use from food chains. Information in terms of this food waste being reused by other industries such as the animal industry is scarce but one can hazard a guess that this is more likely to be low.

Food waste has historically been recycled as livestock feed, particularly for pigs, where cooked food waste (swill) was fed to pigs. As a monogastric species, the pigs' digestive system is well adapted for the conversion of food waste into animal protein (Westendorf, 2000). Swill can be a high-quality animal feed that requires no additional land to be brought into production, and hence has minimal or even positive environmental impact (food waste otherwise disposed). However, the use of swill is prohibited in many countries due to biosecurity and food safety concerns/regulation. In addition, collection and processing costs, nutrient value and even implications on product quality are issues that further hinder the use of food waste as an animal feed ingredient. Hence, for the most part food waste remains a largely untapped resource for the pig industry.

The use of co-products from manufacturing processes (dairy and oilseeds) and surplus human food (bakery products, dairy, confectionary, horticulture, etc.) are, however, well utilised by the pig industry and have been so for quite some time.

The use of distillers dried grains with solubles (DDGS) has been researched for over 50 years. It is now well established for use in just about all pig diets beginning from 2–3 weeks' post weaning in concentrations up to 30% DDGS, through to the lactating and gestating sow up to 30 and 50%, respectively, without any negative effect on growth performance (Stein and Shurson, 2009). The impact on carcass quality, however, is also well documented with an increase in the iodine value of fat (softer fat) from pigs fed DDGS.

There is considerable interest in developing micro- and macroalgae as a sustainable resource that can be used to supply a substantial part of the diet to all phases of pig production. Much of the earlier work has focused on the production of algae as a method of reducing the carbon output of many heavy industries that exhaust significant carbon to the environment. The algal product that is produced as a co-product is available to the animal industries as a feed ingredient. Despite the potential use of algae as a valuable protein source for animals, there has been very little research on the evaluation of algae as a feed ingredient for animals over the past 20-30 years. Lee et al. (1980) fed pigs primarily *Micractinium* sp. and *Scenedesmus* sp. containing ~59% protein, and reported that pig diets containing 8% algae provided comparable performance to corn–soybean meal-fed control groups. Similarly, Taiganides (1992) reported, albeit on a very small study, that steam-boiled algae could replace half of the protein in a pig diet (normally provided by soybean meal) with no significant reduction in growth rate. More recently, Henman (2012) showed that a 10% inclusion of algae meal (as direct replacement for canola meal) in the weaner diet was found to have no negative growth performance effects. As we look to reduce our environmental footprint of pork production, waste management is increasingly being viewed as ways to reduce production costs and generate additional revenue. The ultimate aim here is the exploitation of microalgae for both on-farm wastewater treatment and as a source of feed for pigs. Significant research in Australia is being directed to this very aim. Whilst the challenges of growing algae on on-farm wastewater are significant, the challenges to generate a suitable pig feed ingredient in significant quantities remain a major hurdle. In addition, the food safety and biosecurity risks, similar to those posed by the use of swill in pig production, further complicate the use of such ingredients in pig feed.

The role of a variety of insects that are part of a human diet in many countries is quickly becoming a mainstream but 'micro-niche' protein for human consumption. Insect meal, as it is sometimes referred to by media, may also provide an opportunity to utilise certain food wastes to generate novel pig feed ingredients.

7 Optimising gut health and nutrient utilisation capacity

At a time where antimicrobial resistance (AMR) poses a very real threat to human health, the animal industries, whilst unfairly blamed by media as the sole cause of these issues, do contribute to some of the risk around AMR in pigs, farm workers and the general community. The use of antibiotics as growth promoters was extensive in most pig production systems; however, their continued use irrespective of efficacy is unsustainable. Over the last ten years we are seeing concerted effort to minimise and eventually cease their use globally. Some countries have even gone as far as banning their use in pig production, although the corresponding and significant increase in therapeutic use of antibiotics suggests that perhaps such bans may be less than effective and may take longer for industries to adhere to. Needless to say, the issue of antibiotic use and AMR has caused an increase in both the use and development of alternative nutritional strategies to antibiotic feed additives to improve pig growth performance.

'Gut health' is a term that we define as describing a generalised condition of homeostasis in the gastrointestinal tract (GIT) of the pig (Pluske et al., 2007). As can be expected, the factors and conditions involved in 'gut health' are multifactorial, complex and in many instances poorly described and sometimes incorrectly interpreted, although it is evident that perturbations of the GIT that cause disease, even sub-clinical disease, are a disturbance of this homeostasis. In addition to enteric disease, other factors including the period immediately after weaning, changes in diet and disruptions to meal patterns that disrupt the flow of nutrients through the GIT also affect gut health (Pluske et al., 2007). Gut health can be viewed as an outcome (positive, negative, the status quo) of the complex interactions occurring in the GIT between nutrition (e.g. feed type, feed composition, presence or absence of antibiotic feed additives), the mucosa of the GIT (e.g. type of receptors, mucin, inflammatory activity, cytokine production and barrier function) and the microbiota (e.g. density, location, pathogenicity and extent of colonisation; Pluske et al., 2007). Whilst much of our research focus has been directed to the newly weaned pig, it is important to consider that growing-finishing pigs and sows also suffer from GIT disruptions causing decreased growth performance, morbidity and sometimes death.

In many ways, it is impossible to talk about gut health and growth performance without discussing the role of in-feed antibiotics. In-feed antibiotic additives have been used extensively and successfully in pig diets over many years to improve health and performance. The effect of antibiotic feed additive use in pig production has been variable to say the least, with the relative improvements in growth rate resulting from adding antibiotics to diets for pigs being inversely related to the growth rate of control animals (Braude et al., 1953; cited by Page, 2006). Risks associated predominately with antibiotic resistance of microbes in humans have caused the banning or restriction of antibiotics use as feed additives for pig production in some parts of the world.

8 Understanding nutrition and health interactions

The feed additives/nutritional strategies claiming to alter the microbiota of the GIT to influence 'health' of GIT resulting in improved growth performance have increased since this focus on AMR, and a range of these additives are discussed below.

Zinc oxide (ZnO) is one such product that appears to influence and benefit 'gut health' and growth performance. Numerous studies have shown the production and anti-diarrhoeal benefits of including ZnO at supraphysiological levels (i.e. 2000–3000 mg/kg or ppm) in the diet after weaning, but what is less clear is/are the mechanism(s) whereby ZnO exerts its beneficial effects (Pluske et al., 2007). Reported effects include the increased gene expression of antimicrobial peptides in the small intestine (Wang et al., 2004); positive effects on the stability and diversity of the microflora, particularly with respect to coliforms (Katouli et al., 1999) increased IGF-I and IGF-IR expression in the small intestinal mucosa (Li et al., 2006); bactericidal functions (Jensen-Waern et al., 1998) and reductions in electrolyte secretion *in vitro* from enterocytes (Carlson et al., 2006). Hedemann et al. (2006) found changes in some pancreatic enzymes and mucin staining but concluded that there were no definite answers as to how the growth-promoting and diarrhoea-reducing effects of excess dietary Zn were exerted. Nevertheless, some studies have reported no benefits of feeding ZnO (e.g. Jensen-Waern et al., 1998; Broom et al., 2006), perhaps again a reflection of the 'growth rate of control animals'.

The restrictions on or the banning of in-feed antibiotics has seen an increase in the use of probiotics in the period after weaning, to improve gut health and maintain or improve growth performance. Hillman (2001) showed that intestinal lactobacilli having anti-pathogenic ('probiotic') activity against pathogenic *Escherichia coli* that possessed K88 fimbriae were not evenly distributed across 19 Scottish pig farms. Whilst the need for such probiotics may be warranted, the fact that we are relying on the intake (feed, water) at a time when feed intake is both low and variable (Pluske et al., 1997) certainly has contributed to the success (or not) of such products on improving gut health and growth performance of the young pig. With regard to efficacy of probiotics after weaning, the published data defining the benefits of probiotics for nursery pigs are equivocal, which is not a real surprise given the different species and strains that are used and the wide array of weaning and feeding conditions that products work under (Pluske, 2006).

Work with fermented liquid feed (Demeckova et al., 2002) and *Bacillus cereus* var. *toyoi* (e.g. Taras et al., 2005) suggested that an altered microbiota in the faeces of the dam caused by changing the microbiota of the sow exerts a beneficial influence on both pre- and post-weaning development of the young pig. In the study by Taras et al. (2005), one group of sows were fed for 17 weeks, from Day 24 after mating to Day 28 after farrowing, and the piglets from these sows were fed for six weeks, from Day 15 of lactation to eight weeks of age. The control group of sows/piglets did not receive the probiotic strain. The *Bacillus cereus* var. *toyoi* strain was recovered from the faeces of sows and piglets throughout the trial, including the period 0–14 days of age before introduction of the starter diet occurred, and there was an improvement in FCR of pigs in the post-weaning period derived from sows fed the probiotic during pregnancy and lactation.

Studies by Pluske et al. (2002) and Hampson et al. (2006) showed that feeding a diet low in both soluble non-starch polysaccharides (NSP) and resistant starch (RS) generally afforded protection against *B. hyodysenteriae*. However, the way in which these grains have been processed also appears to be important, especially with cereals inherently low in NSP (<1 g/100 g soluble NSP). The data suggested that a reduction in RS levels (e.g. via extrusion and steam flaking) would only prove effective against SD if the grain in question has an inherently low NSP level to begin with (Durmic et al., 2002). Durmic et al. (2002) reported that colonisation by spirochaetes was highly related to the dietary concentrations of soluble NSP, while development of SD was similarly influenced by the RS content of the diet.

The addition of plasma products to diets has revolutionised the feeding of newly weaned pigs, and especially of pigs weaned before 18 days of age, in the last 25 years. Products such as spray-dried porcine plasma (SDPP), where used, generally enhance performance through increased feed intake and feed efficiency in the immediate post-weaning period. Mechanisms for such effects have been studied (see review by Lallès et al., 2004), including immunoglobulin-independent, glycoprotein-enhanced protection against *E. coli* and the specific protection brought about by plasma immunoglobulins (van Dijk et al., 2001). Spray-dried plasma protein was shown to lower the levels of gene expression of inflammatory cytokines after an Enterotoxic Escherichia coli (ETEC) challenge (Bosi et al., 2004).

Numerous reviews have been written regarding the use of organic and inorganic acids in pig diets (e.g. Partanen, 2001; Decuypere and Dierick, 2003; Mroz et al., 2006). The positive effects of feeding acids to pigs on gut health and development, and indirectly on pig health and productivity, may be attributed to various factors. These include (1) antimicrobial activity of non-dissociated organic acids; (2) lowering digesta pH, in particular in the stomach, aiding protein digestion; (3) lowering stomach emptying rate; (4) stimulating (pancreatic) enzyme production and activity in the small intestine; and (5) providing nutrients that are preferred by intestinal tissue thereby enhancing mucosal integrity and function. Because of these beneficial and synergistic effects, different combinations of organic and inorganic acids are used widely in diets for newly weaned pigs, and used increasingly in diets for growing-finishing pigs and sows. The effectiveness of feeding acids to pigs will vary with the types and combinations of acid, the animal's state and feed characteristics, in particular the diet's buffering capacity (Blank et al., 1999; Mroz et al., 2006). A relatively recent development is the encapsulation of acids for targeted delivery to different gut segments. Studies such as those conducted by Piva et al. (2007) have shown that relatively simple encapsulation is effective in delaying absorption of dietary acids and allowing more effective delivery of acids to the distal ileum, caecum and colon of piglets.

Halas et al. (2009) tested the hypothesis that feeding diets supplemented with inulin (8% of the diet) and benzoic acid (0.5% of the diet) will reduce faecal ETEC shedding and the incidence of Post-weaning diarrhea (PWD), following oral dosing of pigs after weaning with 3 ml of broth containing 3×10^7, 2×10^9, 1×10^{10} and 5×10^8 colony-forming units (CFU) of a freshly grown strain of ETEC serotype O149:K91:K88 (toxins LT, STa, and STb). Dietary supplementation with inulin, either alone or in combination with benzoic acid, reduced the number of days of diarrhoea in pigs weaned at 21 days of age without significantly affecting average ETEC shedding.

Phosphorylated mannans derived from yeast (*Saccharomyces cerevisiae*) cell wall-stimulated phagocytosis of intestinal macrophages and increased the proportion of CD8+ T lymphocytes in the jejunal lamina propria, in addition to improving growth performance (Davis et al., 2004). The percentages of neutrophils and lymphocytes in plasma were lower and higher, respectively, in the supplemented pigs, suggesting a lower level of stress with mannan in these pigs. In another study, however, plasma haptoglobin was higher in the mannan-supplemented pigs (Burkey et al., 2004). Diet supplementation with live yeast also increased the numbers of macrophages in the ileum (Bontempo et al., 2006).

Carver and Walker (1995) extensively reviewed the role of nucleotides in human nutrition. The potential beneficial effects to the immune system, small intestine growth and development, lipid metabolism and hepatic function highlighted the potential application to animal nutrition. The inclusion of supplemental nucleotides in livestock diets for similar benefits is far from novel; however, using yeast extract as the delivery vehicle has only come about in the last 10-15 years.

Amino acids have come into consideration for specific usage as our ability to measure their influence has increased. Burrin and Stoll (2003) reviewed a number of studies with reference to the amino acid arginine and suggested that given the considerable advances made in the understanding of intestinal nutrient utilisation and metabolism, an alternate or even complimentary goal in nutrition might be to formulate young pig diets with the specific task of optimising the growth, function and health of the gut.

Glutamine is an abundant amino acid found in plant and animal proteins and due to its abundance, it has not traditionally been considered a nutrient needed in livestock diets. However, advances in analytical methods have shown the abundance of glutamine within the fluids and tissues of the body, suggesting that it may be more important than we have traditionally thought, especially for the young piglet. A recent review by Wu et al. (2011) highlighted the important roles of glutamine. This information has led to various studies being conducted in the weaned pig. Johnson et al. (2006) fed 21-day-old weaned pigs diets containing an additional 4.38% glutamine for 14 days. Whilst no effect was seen on feed intake or final weight, they saw significant modifications, potentially beneficial, of immune cells in mesenteric lymph nodes. As a result of these studies and work of their own, Wu et al. (2011) have suggested that specific diets should be supplemented with 1% glutamine. A gestation diet fed from Day 90 onwards would ameliorate foetal growth retardation and pre-weaning mortality. A lactation diet containing 1% glutamine would enhance milk production, and in creep diets it would help maintain gut health, prevent intestinal dysfunction, and increase growth and survival.

The growth performance benefits of products containing yeast extract and peptides to replace plasma in the young pigs have been previously summarised (Tibbetts, 2000). Mahan (1999) demonstrated that peptide protein products in combination with yeast extract could completely replace plasma protein if diets were correctly formulated. In diets not containing blood plasma, these vegetable proteins have been shown to improve feed intake against controls and other animal protein sources (Maribo, 2001). Carlson et al. (2001) conducted a study to evaluate the effect of plasma and yeast extract protein supplementation in the young pigs on lifetime performance. Overall average daily gain during the finisher period and slaughter weight for pigs from the yeast extract protein nursery group was significantly higher compared to gains of pigs from either of the control and plasma treatment groups. Since then, numerous studies have been conducted to further demonstrate the growth performance benefits of yeast extract protein.

9 Future trends and conclusion

In looking back, our understanding of the nutritional requirements and metabolism of the pig has been significantly advanced over the last 10–20 years. From the way we formulate diets, our understanding of the feed ingredients and the nutritional technologies have enabled us to improve the performance of the pig. The magnitude of progress is evident when we compare the performance of pigs today compared to late 1980s; whole of life average daily gain has increased from 450 g/day to >720 g/day, whilst herd feed conversion has decreased from >5.5 to 3.7 kg feed/kg carcass. During this time, significant technologies such as pST and ractopamine that have been key contributors to this improved performance are no longer viable technologies for use by the pork industry. Some of these technologies whilst still effective are no longer viable due to shifting consumer perceptions.

The risk associated with AMR in humans and pigs has seen the use of in-feed antibiotics becoming increasingly unsustainable. This has given rise to a range of new technologies that have positively contributed to both growth performance and animal health. The interaction between nutrition and health is especially apparent in the young pig where technologies to alleviate the growth check is seen as a consequence of weaning.

As we look forward, pig production systems are more than likely to undergo significant change, if the last 10–20 years are anything to go by. Changes in genetics, housing systems, animal health care, as well as systems shaped by changing environmental regulation and consumer attitudes are then likely to affect the nutrient requirements of the modern pig. We will need to pay particular attention to the nutritional requirements of the pig. Our reliance on theoretical data rather than actual nutrient requirement data means we are particularly at risk of 'assuming' we are feeding pigs to grow as closely possible to their genetic potential. The same applies for our estimation of the nutrient specification of feed ingredients. Technologies such as NIR and other 'precision farming' technologies must be better utilised to ensure that the diets we are feeding our pigs actually meet their requirement. Whilst feeder design and innovation to enable us to tailor diets to meet individual pig's requirements, still a distant possibility, does present us with opportunities to advance our understanding of the pigs' nutritional requirements and metabolism.

10 Where to look for further information

This chapter has highlighted the areas within pig nutritional requirements and metabolism that have the potential to impact the sustainability of pig production. When considering these areas, it is imperative to concentrate on the primary drivers of sustainability as outlined by groups like the FAO in documents such as *Building a Common Vision for Sustainable Food and Agriculture - Principles and Approaches* (FAO, 2014).

Our knowledge of nutritional requirements and metabolism continues to evolve, aided by advances in technology and methodology; therefore the most up-to-date information will continue to be found in the scientific journals. A range of references were reviewed in the writing of this chapter and these papers would be an ideal starting point to further understand this topic. In particular, *Metabolic modifiers as performance-enhancing technologies for livestock production* (Dunshea et al., 2016) is a good review of the impact of metabolic modifiers on production, their role in sustainable production and the issues affecting their wider use. *Advances in Pork Production* is the proceedings of the Banff Pork Seminar and has provided a number of references used in this chapter. Covering a range of topics the technology transfer seminars invited papers and abstracts, generally, of high value and have added considerably to our understanding of pig production.

11 References

Alexandratos, N. and Bruinsma, J. (2012). World agriculture towards 2030/2050: the 2012 revision. ESA Working paper No. 12-03. Rome, FAO.

Apple, J. K., Rincker, P. J., McKeith, F. K., Carr, S. N., Armstrong, T. A. and Matzat, P. D. (2007). Meta-analysis of the ractopamine response in finishing swine. *Prof. Anim. Sci.* 23, 179-96.

Ball, R. O., Samuel, R. S. and Moehn, S. (2008). Nutrient requirements of prolific sows. *Adv. Pork Prod.* 19, 223-36.

Bauman, D. E. and Dunshea, F. R. 2010. Somatropin. In Pond, W. G. and Bell, A. W. (Eds), *Encyclopedia of Animal Science.* Second Edition, pp. 995-8. Taylor and Francis: London, UK.

Beltranena, E., Patterson, J., Foxcroft, G. and Pettitt, M. (2005). 12 Non-negotiable aspects of gilt development Western Hog Journal Summer 2004, pp. 45-50.

Bergsma, R., Kanis, E., Verstegen, M. W. A., van der Peet-Schwering, C. M. C. and Knol, E. F. (2009). Lactation efficiency as a result of body composition dynamics and feed intake in sows. *Livest. Sci.* 125, 208-22.

Black, J. L., Flinn, P. C., Diffey, S. and Tredrea, A. M. (2013). An update on near infrared reflectance analysis of cereal grains to estimate digestible energy content for pigs. In Pluske, J. R. and Pluske J. M. (Eds), *Manipulating Pig Production XIV*, p. 54. Australasian Pig Science Association: Werribee, VIC, Australia.

Blank, R., Mosenthin, R., Sauer, W. C. and Huang, S. (1999). Effect of fumaric acid and dietary buffering capacity on ileal and fecal amino acid digestibilities in early-weaned pigs. *J. Anim. Sci.* 77, 2974-84.

Bonneau, M. and Enright, W. J. (1995) Immunocastration in cattle and pigs. *Livest. Prod. Sci.* 42, 193-200 doi: (10.1016/0301-6226(95)00020-I).

Bontempo, V., Di Giancamillo, A., Savoini, G., Dell'Orto, V. and Domeneghini, C. (2006) Live yeast dietary supplementation acts upon intestinal morpho-functional aspects and growth in weanling piglets. *Anim. Feed Sci. Technol.* 129, 224-36.

Bosi, P., Casini, L., Finamore, A., Cremokolini, C., Merialdi, G., Trevisi, P., Nobili, F. and Mengheri, E. (2004). Spray-dried plasma improves growth performance and reduces inflammatory status of weaned pigs challenged with enterotoxigenic Escherichia coli K88. *J. Anim. Sci.* 82, 1764-72.

Braude, R., Wallace, H. D. and Cunha, T. J. (1953). The value of antibiotics in the nutrition of swine. A review. *Antibiot. Chemother.* 3, 271-91.

Brooks, P. H., Smith, D. A. and Irwin, V. C. R. (1977). Biotin supplementation of diets; the incidence of foot lesions, and the reproductive performance of sows. *Vet. Rec.* 101, 46-50.

Broom, L. J., Miller, H. M., Kerr, K. G. and Knapp, J. S. (2006). Effects of zinc oxide and Enterococcus faecium SF68 dietary supplementation on the performance, intestinal microbiota and immune status of weaned piglets. *Res. Vet. Sci.* 80, 45-54.

Burkey, T. E., Dritz, S. S., Nietfeld, J. C., Johnson, B. J. and Minton, J. E. (2004). Effect of dietary mannanoligosaccharide and sodium chlorate on the growth performance, acute-phase response, and bacterial shedding of weaned pigs challenged with Salmonella enterica serotype Typhimurium. *J. Anim. Sci.* 82, 397-404.

Burrin, D. and Stoll, B. (2003). Intestinal nutrient requirements in weanling pigs. In Pluske, J. R., Verstegen, M. W. A. and Le Dividich, H. (Eds), *The Weaner Pig: Concepts and Consequences*, pp. 301-35. Wageningen Academic Publishers Wageningen,: The Netherlands.

Cadogan, D. (2009). Net energy defines lean and fat deposition better than digestible energy. Final Report, Project 2G-104. The Co-operative Research Centre for High Integrity Australian Pork, Roseworthy, SA, Australia.

Campbell, R. G., Steele, N. C., Caperna, T. J., McMurtry, J. P., Solomon, M. B. and Mitchell, A. D. (1988). Interrelationships between energy intake and exogenous porcine growth hormone administration on the performance, body composition and protein and energy metabolism of growing pigs weighing 25 to 55 kilograms body weight. *J. Anim. Sci.* 66, 1643-55.

Campbell, R. G., Steele, N. C., Caperna, T. J., McMurtry, J. P., Solomon, M. B. and Mitchell, A. D. (1989). Effects of exogenous porcine growth hormone administration between 30 and 60 kilograms on the subsequent and overall performance of pigs grown to 90 kilograms. *J. Anim. Sci.* 67, 1265-71.

Carlson, C. S., Hilley, H. D., Meuten, D. J., Hagan, J. M. and Moser, R. L. (1988). Effect of reduced growth rate on the prevalence and severity of osteochondrosis in gilts. *Am. J. Vet. Res.* 49, 396-402.

Carlson, D., Sehested, J. and Poulsen, H. D. (2006). Zinc reduces the electrophysiological responses in vitro to basolateral receptor mediated secretagogues in piglet small intestinal epithelium. *Comp. Biochem. Physiol. A Mol. Integr. Physiol.* 144, 514-19.

Carlson, M. S., Veum, T. L., Turk, J. R., Bollinger, D. W. and Tibbetts, G. W. (2001). A comparison between feeding either peptide or plasma proteins with or without a feed grade antibiotic on pig growth performance and intestinal health. Unpublished. University of Missouri, Columbia, MO, USA.

Carver, J. D. and Walker, W. A. (1995). The role of nucleotides in human nutrition. *J. Nutr. Biochem.* 6, 58-72.

Chung, C. S., Etherton, T. D. and Wiggins, J. P. (1985). Stimulation of swine growth by porcine growth hormone. *J. Anim. Sci.* 60, 118-30.

Clowes, E. J., Aherne, F. X., Foxcroft, G. R. and Baracos, V. E. (2003). Selective protein loss in lactating sows is associated with reduced litter growth and ovarian function. *J. Anim. Sci.* 81, 753-64.

Cools, A., Maes, D., Decaulwe, R., Buyse, J., van Kempen, T. A. T. G., Liesegang, A. and Janssens, G. P. J. (2014). Ad libitum feeding during the peripartal period affects body condition, reproduction results and metabolism of sows. *Anim. Reprod. Sci.* 145, 130-40.

Cronin, G. M., Dunshea, F. R., Butler, K. L., McCauley, I., Barnett, J. L. and Hemsworth, P. H. (2003). The effects of immuno- and surgical-castration on the behaviour and consequently growth of group-housed, male finisher pigs. *Appl. Anim. Behav. Sci.* 81, 111-26.

Davis, M. E., Maxwell, C. V., Erf, G. F., Brown, D. C. and Wistuba, T. J. (2004). Dietary supplementation with phosphorylated mannans improves growth response and modulates immune function of weanling pigs. *J. Anim. Sci.* 82, 1882-91.

De Bettio, S., Majorka, A., Barrilli, L. N. E., Bergsma, R. and Silva, B. A. N. (2016). Impact of feed restriction on the performance of highly prolific lactating sows and its effect on the subsequent lactation. *Animal* 10, 396-402.

de Lange, C. F. M. and Birkett, S. H. (2005). Characterization of useful energy contents in swine and poultry feed ingredients. *Can. J. Anim. Sci.* 85, 269-80.

de Lange, C. F. M. (2013). New NRC (2012) nutrient requirements of swine. *Adv. Pork Prod.* 24, 17-28.

De Moraes, P. J. U., Allison, J., Robinson, J. A., Baldo, G. L., Boeri, F. and Borla, P. (2013). Life cycle assessment (LCA) and environmental product declaration (EPD) of an immunological product for boar taint control in male pigs. *J. Environ. Assess. Policy Manag.* 15, 1350001-1350026.

Decuypere, J. A. and Dierick, N. A. (2003). The combined use of triacylglycerols containing medium-chain fatty acids and exogenous lipolytic enzymes as an alternative to in-feed antibiotics in piglets: concept, possibilities and limitation- An overview. *Nutr. Res. Rev.* 16, 193-209.

Demeckova, V., Kelly, D., Coutts, A. G. P., Brooks, P. H. and Campbell, A. (2002). The effect of fermented liquid feeding on the faecal microbiology and colostrum quality of farrowing sows. *Int. J. Food Microbiol.* 79, 85-97.

Donovan, T. S. and Dritz, S. S. (2000). Effect of split nursing on variation in pig growth from birth to weaning. *J. Am. Vet. Med. Assoc.* 217, 79-81.

Douglas, S. L., Edwards, S. A. and Kyriazakis, I. (2014). Management strategies to improve the performance of low birth weight pigs to weaning and their long-term consequences. *J. Anim. Sci.* 92, 2280-8.

Douglass, G. K. (1984). The meanings of agricultural sustainability. In Douglass, G. K. (Ed.), *Agricultural Sustainability in a Changing World Order*, pp. 3-29. Westview Press: Boulder, CO, USA.

Dowling, H. J., Offenbacher, E. G. and Pi-Sunyer, F. X. (1989). Absorption of inorganic, trivalent chromium from the vascularly perfused rat small intestine. *J. Nutr.* 119, 1138-45.

Ducros, V. (1992). Chromium metabolism. A literature review. *Biol. Trace Elem. Res.* 32, 65-77.

Dunshea, F. R. (1993). Effect of metabolism modifiers on lipid metabolism in the pig. *J. Anim. Sci.* 71, 1966-77.

Dunshea, F. R. (2005). Sex and porcine somatotropin impact on variation in growth performance and back fat thickness. *Aust. J. Exp. Agr.* 45, 677-82.

Dunshea, F. R., Allison, J. R. D., Bertram, M., Boler, D. D., Brossard, L., Campbell, R., Crane, J. P., Hennessy, D. P., Huber, L., de Lange, C., Ferguson, N., Matzat, P., McKeity, F., Moraes, P. J. U., Mullan, B. P., Noblet, J., Quiniou, N. and Tokach, M. (2013). The effect of immunization against GnRF on nutrient requirements of male pigs: a review. *Animal* 7, 1769-78.

Dunshea, F. R., Colantoni, C., Howard, K., McCauley, I., Jackson, P., Long, K. A., Lopaticki, S., Nugent, E. A., Simons, J. A., Walker, J. and Hennessy, D. P. (2001). Vaccination of boars with a GnRH vaccine (Improvac) eliminates boar taint and increases growth performance. *J. Anim. Sci.* 79, 2524-35.

Dunshea, F. R., D'Souza, D. N. and Channon, H. A. (2016). Metabolic modifiers as performance-enhancing technologies for livestock production. *Anim. Front.* 6, 6-14.

Dunshea, F. R., D'Souza, D. N., Pethick, D. W., Harper, G. S. and Warner, R. D. (2005). Effects of dietary factors and other metabolic modifiers on quality and nutritional value of meat. *Meat Sci.* 71, 8-38.

Durmic, Z., Pethick, D. W., Mullan, B. P., Accioly, J. M., Schulze, H. and Hampson, D. J. (2002). Evaluation of large-intestinal parameters associated with dietary treatments designed to reduce the occurrence of swine dysentery. *Br. J. Nutr.* 88, 159-69.

Etherton, T. D., Wiggins, J. P., Evock, C. M., Chung, C. S., Rebhun, J. F., Walton, P. E. and Steele, N. C. (1987). Stimulation of pig growth performance by porcine growth hormone: determination of the dose-response relationship. *J. Anim. Sci.* 64, 433-43.

Fairlie, S. (2010). *Meat: A Benign Extravagance*. Permanent Publications: East Meon UK,.

FAO (2013). *Food Wastage Footprint - Impacts on Natural Resources*. Food and Agriculture Organization of the United Nations: Rome, Italy.

FAO (2014). *Building a Common Vision for Sustainable Food and Agriculture - Principles and Approaches*. Food and Agriculture Organization of the United Nations: Rome, Italy.

Foley, J. A., Ramankutty, N., Brauman, K. A., Cassidy, E. S., Gerner, J. S., Johnston, M., Mueller, N. D., O'Connell, C., Ray, D. K., West, P. C., Balzer, C., Bennett, E. M., Carpenter, S. R., Hill, J., Monfreda, C., Polasky, S., Rockstrom, J., Sheehan, J., Siebert, S., Tilman, D. and Zaks, D. P. M. (2011). Solutions for a cultivated planet. *Nature* 478, 337-42.

Foxcroft, G. R. (1997). Mechanisms mediating nutritional effects on embryonic survival in pigs. *J. Reprod. Fertil.* 52 (Supp.), 47-61.

Foxcroft, G. R. (2007). Pre-natal programming of variation in post-natal performance: how and when? *Adv. Pork Prod.* 18, 167-89.

Foxcroft, G. R. and Town, S. (2004). Prenatal programming of postnatal performance: the unseen cause of variance. *Adv. Pork Prod.* 15, 269-79.

Gerber, P. J., Steinfeld, H., Henderson, B., Mottet, A., Opio, C., Dijkman, J., Falcucci, A. and Tempio, G. (2013). *Tackling Climate Change Through Livestock - A Global Assessment of Emissions and Mitigation Opportunities*. Food and Agriculture Organization of the United Nations (FAO): Rome.

Giles, D. D. (1942). An experiment to determine the effect of growth hormone of the anterior lobe of the pituitary gland on swine. *Am. J. Vet. Res.* 3, 77-86.

Greer, E. B., Leibholz, J. M., Pickering, D. I., Macoun, R. E. and Bryden, W. L. (1991). Effect of supplementary biotin on the reproductive performance, body condition and foot health of sows on three farms. *Aust. J. Agr. Res.* 42, 1013-21.

Gresham, A. (2003). Infectious reproductive disease in pigs. *In Practice* 25, 466-73.

Halas, D., Hansen, C. F., Hampson, D. J., Mullan, B. P., Wilson, R. H. and Pluske, J. R. (2009). Effect of dietary supplementation with inulin and/or benzoic acid on the incidence and severity of post-weaning diarrhoea in weaner pigs after experimental challenge with enterotoxigenic Escherichia coli. *Arch. Anim. Nutr.* 63, 267-80.

Hamilton, C. R. and Veum, T. L. (1984). Response of sows and litters to added biotin in environmentally regulated facilities. *J. Anim. Sci.* 59, 151-7.

Hampson, D. J., Fellstrom, C. and Thomson, J. R. (2006). Swine dysentery. In Straw, B. E., Zimmerman, J. J., D'Allaire, S. and Taylor, D. J. (Eds), *Diseases of Swine*, pp. 785-805. Blackwell Publishing: Iowa, USA.

Hedemann, M. S., Jensen, B. B. and Poulsen, H. D. (2006). Influence of dietary zinc and copper on digestive enzyme activity and intestinal morphology in weaned pigs. *J. Anim. Sci.* 84: 3310-20.

Henman, D. J. (2012). Evaluation of Algal meal as an energy and protein source in pig diets. Final Report, Project 4A-102. The Co-operative Research Centre for High Integrity Australian Pork, Roseworthy, SA, Australia.

Hewitt, R. J. E., Corso, A. and van Barneveld, R. J. (2015). Reducing variation in finisher growth performance through early post-weaning dietary intervention. *Anim. Prod. Sci.* 55, 1573.

Hillman, K. (2001). Bacteriological aspects of the use of antibiotics and their alternatives in the feed of non-ruminant animals. In Garnsworthy, P. C. and Wiseman, J. (Eds),

Recent Advances in Animal Nutrition, pp. 107–34. Nottingham University Press: Loughborough, UK.

Hoving, L. L., Soede, N. M., Feitsma, H. and Kemp, B. (2012). Embryo survival, progesterone profiles and metabolic responses to an increased feeding level during second gestation in sows. *Theriogenology 77*, 1557–69.

Hung, A. T., Leury, B. J., Sabin, M. A., Lien, T. F. and Dunshea, F. R. (2015). Dietary chromium picolinate of varying particle size improves carcass characteristics and insulin sensitivity in finishing pigs fed low- and high-fat diets. *Anim. Prod. Sci.* 55, 454–60.

Jensen-Waern, M., Melin, L., Lindberg, R., Johannisson, A., Petersson, L. and Wallgren, P. (1998). Dietary zinc oxide in weaned pigs-effects on performance, tissue concentrations, morphology, neutrophil functions and faecal microflora. *Res. Vet. Sci.* 64, 225–31.

Johnson, I. R., Ball, R. O., Baracos, V. E. and Field, C. J. (2006). Glutamine supplementation influences immune development in the newly weaned piglet. *Dev. Comp. Immunol.* 30, 1191–202.

Katouli, M., Melin, L., Jensen-Waern, M., Wallgren, P. and Mollby, R. (1999). The effect of zinc oxide supplementation on the stability of the intestinal flora with special reference to composition of coliforms in weaned pigs. *J. Appl. Microbiol.* 87, 564–73.

Kil, D. Y., Kim, B. G. and Stein, H. H. (2013). Feed energy evaluation for growing pigs. Asian-Australasian *J. Anim. Sci.* 26, 1205–17.

Kim, J. C. and Mullan, B. P. (2012). Quantification of the variability in the amino acid and reactive lysine content of soybean meal and development of a NIR calibration for rapid prediction of reactive lysine content. Final Report, Project 4B-106. The Co-operative Research Centre for High Integrity Australian Pork, Roseworthy, SA, Australia.

King, R. H., Campbell, R. G., Smits, R. J., Morley, W. C., Ronnfeldt, K., Butler, K. and Dunshea, F. R. (2000). Interrelationships between dietary lysine, sex, and porcine somatotropin administration on growth performance and protein deposition in pigs between 80 and 120 kg live weight. *J. Anim. Sci.* 78, 2639–51.

Knauer, M. T. (2016). Effect of increasing sow feeding level in late gestation on piglet quality and sow body condition. *J. Anim. Sci.* 94 (Supp 2), 97.

Koketsu, Y., Dial, G. D., Pettigrew, J. E., Marsh, W. E. and King, V. L. (1996). Characterization of feed intake patterns during lactation in commercial swine herds. *J. Anim. Sci.* 74, 1202–10.

Kopinski, J. S. and Leibholz, J. (1989) Biotin studies in pigs. 2. *Br. J. Nutr.* 62, 761–6.

Kopinski, J. S., Leibholz, J., Bryden, W. L. and Fogarty, A. C. (1989). Biotin studies in pigs. 1. *Br. J. Nut.* 62, 751–9.

Kornegay, E. T. (1986). Biotin in swine production: A review. *Livest. Prod. Sci.* 14, 65–89.

Lalles, J. P., Boudry, G., Favier, C., Le Floc'h, N., Lurona, I., Montagne, L., Oswald, I. P., Pie, S., Piel, C. and Seve, B. (2004). Gut function and dysfunction in young pigs: Physiology. *Anim. Res.* 4, 301–16.

Lawlor, P. G., Lynch, P. B., O'Connell, M. K., McNamara, L., Reid, P. and Stickland, N. C. (2007). The influence of over feeding sows during gestation on reproductive performance and pig growth to slaughter. *Arch. Anim. Breed.* 50, 82–91.

Lee, B. Y., Lee, K. W., McGarry, M. G. and Graham, M. (1980). *Treatment and Resource Recovery.* Report of a Workshop on High-Rate Algae Ponds, 27–29 February 1980. Singapore, p. 27. IDRC, Canada.

Lewis, A. J., Cromwell, G. L. and Pettigrew, J. E. (1991). Effects of supplemental biotin during gestation and lactation on reproductive performance of sows: a cooperative study. *J. Anim. Sci.* 69, 207–14.

Li, X. L., Yin, J. D., Li, D. F., Chen, X. J., Zang, J. J. and Zhou, X. (2006). Dietary supplementation with zinc oxide increases IGF-I and IGF-I receptor gene expression in the small intestine of weanling piglets. *J. Nutr.* 136, 1786–91.

Mahan, D. C. (1999). Comparison of plasma protein and Ultimate Protein in the diets of starter pigs. Report to Alltech. The Ohio State University, Columbus, OH, USA.

Mahan, D. C. and Newton, E. A. (1995). Effect of initial breeding weight on macro- and micromineral composition over a three=parity period using a high-producing sow genotype. *J. Anim. Sci.* 73: 151–8.

Maribo, H. (2001). Commercial products for weaners: NuPro 2000 as an alternative protein source for weaners. Report no 256. The National Committee for Pig Production, Danish Bacon and Meat Council, Danske Slagterier, Denmark.

McCauley, I., Watt, M., Suster, D., Kerton, D. J., Oliver, W. T., Harrell, R. J. and Dunshea, F. R. (2003). A GnRF vaccine (Improvac ®) and porcine somatotropin (Reporcin ®) have synergistic effects upon growth performance in both boars and gilts. *Aust. J. Agr. Res.* 54, 11–20.

Mertz, W. (1969). Chromium occurrence and function in biological systems. *Physiol. Rev.* 49, 163–239.

Moore, K. L., Dunshea, F. R., Mullan, B. P., Hennessy, D. P. and D'Souza, D. N. (2009) Ractopamine supplementation increases lean deposition in entire and immunocastrated male pigs. *Anim. Prod. Sci.* 49, 1113–19.

Mroz, Z., Koopmans, S. J., Bannink, A., Partanen, A. K., Krasucki, W., Overland, M. and Radcliffe, S. (2006). Carboxylic acids as bioregulators and gut growth promoters in non-ruminants. In Mosenthin, R., Zentek, J. and Zebrowska, T. (Eds), *Biology of Nutrition in Growing Animals*, pp. 81-133. Elsevier Limited: Amsterdam, The Netherlands.

Nakano, T., Brennan, J. and Aherne, F. (1987). Leg weakness and osteochondritis in swine: a review. *Can. J. Anim. Sci.* 67, 883–901.

National Research Council (1998). *Nutritional Requirements of Swine*. 10th edition. National Academies Press:, Washington, DC USA.

National Research Council (2012). *Nutritional Requirements of Swine*. 11th edition. National Academies Press: Washington, DC, USA.

Nitrayova, S., Heger, J., Patras, P. and Brestensky, M. (2006). Effect of body weight and pig individuality on apparent ileal digestibility of amino acids and total nitrogen. *Slovac J. Anim. Sci.* 39, 65–8.

Noblet, J. and Shi, X. S. (1993). Comparative digestibility of energy and nutrients in growing pigs fed ad libitum and adult sows fed at maintenance. *Livest. Prod. Sci.* 34, 137–52.

Noblet, J. (2006). Recent advances in energy evaluation of feeds for pigs. In Garnsworthy, P. C. and Wiseman, J. (Eds), *Recent Advances in Animal Nutrition*, pp. 1-26. Nottingham University Press: Nottingham, UK.

O'Connell, N. E., Beattie, V. E. and Watt, D. (2005). Influence of regrouping strategy on performance, behavior and carcass parameters in pigs. *Livest. Prod. Sci.* 97, 107–15 doi: (10.1016/j.livprodsci.2005.03.005).

Oliver, W. T., McCauley, I., Harrell, R. J., Suster, D., Kerton, D. J. and Dunshea, F. R. (2003). A gonadotropin-releasing factor vaccine (Improvac) and porcine somatotropin have synergistic and additive effects on growth performance in group-housed boars and gilts. *J. Anim. Sci.* 81, 1959–66.

Page, S. W. (2006). Current use of antimicrobial growth promoters in food animals: The benefits. In Barug, D., de Jong, J., Kies, A. K. and Verstegen, M. (Eds), *Antimicrobial Growth Promoters: Where Do We Go from Here?*, pp. 19–51. Wageningen Academic Publishers: Wageningen, The Netherlands.

Partanen, K. (2001). Organic acids – Their efficacy and modes of action in pigs. In Piva, A., Bach Knudsen, K. E. and Lindberg, J. E. (Eds), *Gut Environment of Pigs*, pp. 201–18. Nottingham University Press: Loughborough, UK.

Patience, J. F., Engele, K., Beaulieu, A. D., Gonyou, H. W. and Zijlstra, R. T. (2004). Variation: costs and consequences. *Adv. Pork Prod.* 15, 257–66.

Penny, R. H. C., Cameron, R. D. A., Johnson, S., Kenyon, P. J., Smith, H. A., Bell, A. W. P., Cole, J. P. L. and Taylor, J. (1981). Influence of biotin supplementation on sow reproductive efficiency. *Vet. Rec.* 109, 80–1.

Piva, A., Pizzamiglio, V., Morlacchini, M., Tedeschi, M. and Piva, G. (2007). Lipid microencapsulation allows slow release of organic acids and natural identical flavors along the swine intestine. *J. Anim. Sci.* 85, 486–93.

Pluske, J. R. (2006). New thoughts on nutrition of newly weaned pigs. In Murphy, J. M. and Kane, T. M. (Eds), Proceedings of the 6th London Swine Conference, pp. 67–77. London Swine Conference: Ontario, Canada.

Pluske, J. R., Hansen, C. F., Payne, H. G., Mullan, B. P., Kim, J. C. and Hampson, D. J. (2007). Gut health in the pig. In Paterson, J. E. and Barker, J. A. (Eds), *Manipulating Pig Production*. Volume XI., pp. 147–58. Australasian Pig Science Association: Werribee,VIC, Australia.

Pluske, J. R., Pethick, D. W., Hopwood, D. E. and Hampson, D. J. (2002). Nutritional influences on some major enteric bacterial diseases of pigs. *Nutr. Res. Rev.* 15, 333–71.

Pluske, J. R., Williams, I. H. and Hampson, D. J. (1997). Factors influencing the structure and function of the small intestine in the weaned pig: a review. *Livest. Prod. Sci.* 51, 215–36.

Quesnel, H., Brossard, L., Valancogne, A. and Quiniou, N. (2008). Influence of some sow characteristics on within-litter variation of piglet birth weight. *Animal* 2, 1842–9.

Quiniou, N., Dagorn, J. and Gaudré, D. (2002). Variation of piglets' birth weight and consequences on subsequent performance. *Livest. Prod. Sci.* 78, 63–70.

Rehfeldt, C., Kuhn, G., Nürnberg, G., Kanitz, E., Schneider, F., Beyer, M., Nürnberg, K. and Ender, K. (2001). Effects if exogenous somatotropin during early gestation on maternal performance, fetal growth, and compositional traits in pigs. *J. Anim. Sci.* 79, 1789–99.

Revell, D. K., Williams, I. H., Mullan, B. P., Ranford, J. L. and Smits, R. J. (1998a). Body composition at farrowing and nutrition during lactation affect the performance of primiparous sows: I. voluntary feed intake, weight loss, and plasma metabolites. *J. Anim. Sci.* 76, 1729–37.

Revell, D. K., Williams, I. H., Mullan, B. P., Ranford, J. L. and Smits, R. J. (1998b). Body composition at farrowing and nutrition during lactation affect the performance of primiparous sows: II. milk composition, milk yield, and pig growth. *J. Anim. Sci.* 76, 1738-43.

Rikard-Bell, C., Curtis, M. A., van Barneveld, R. J., Mullan, B. P., Edwards, A. C., Gannon, N. J., Henman, D. J., Hughes, P. E. and Dunshea, F. R. (2009). Ractopamine hydrochloride improves growth performance and carcass composition in immunocastrated boars, intact boars, and gilts. *J. Anim. Sci.* 87, 3536-43.

Rutherfurd, S. M., Moughan, P. J. and Morel, P. C. H. (1997). Assessment of the true ileal digestibility of reactive lysine as a predictor of lysine uptake from the small intestine of the growing pig. *J. Agr. Food Chem.* 45, 4378-83.

Sales, J. (2011). A meta-analysis of the effects of dietary betaine supplementation on finishing performance and carcass characteristics of pigs. *Anim. Feed Sci. Technol.* 165, 68-78.

Sales, J. and Jancik, F. (2011). Effects of dietary chromium supplementation on performance, carcass characteristics, and meat quality of growing-finisher swine: a meta-analysis. *J. Anim. Sci.* 89, 4054-67.

Schrama, J. W., Heetkamp, M. J., Simmins, P. H. and Gerrits, W. J. (2003). Dietary betaine supplementation affects energy metabolism of pigs. *J. Anim. Sci.* 81, 1202-9.

Steele, N. C., Althen, T. G. and Frobish, L. T. (1977). Biological activity of glucose tolerance factor in swine. *J. Anim. Sci.* 45, 1341-5.

Stein, H. H. and Shurson, G. C. (2009). The use and application of distillers dried grains with solubles in swine diets. *J. Anim. Sci.* 87, 1292-303.

Sugiyama, T., Kusuhara, S., Chung, T. K., Yonekura, H., Azem, E. and Hayakama, T. (2013). Effects of 25-hydroxy-cholecalciferol on the development of osteochondrosis in swine. *Anim. Sci. J.* 84, 341-149.

Suster, D., Leury, B. J., King, R. H., Mottram, M. and Dunshea, F. R. (2004). Interrelationships between porcine somatotropin (pST), betaine, and energy level on body composition and tissue distribution of finisher boars. *Aust. J. Agr. Res.* 55, 983-90.

Taiganides, E. P. (1992). *Pig Waste Management and Recycling: The Singapore Experience.* International Development Research Centre: Ottawa, Canada.

Taras, D., Vahjen, W., Macha, M. and Simon. O. (2005). Response of performance characteristics and fecal consistency to long-lasting dietary supplementation with the probiotic strain Bacillus cereus var. toyoi to sows and piglets. *Arch. Anim. Nutr.* 59, 405-17.

Tibbetts, G. W. (2000). Biopeptides in post-weaning diets for pigs: results to date. In Lyons T. P. and Jacques K.A. (Eds), *Biotechnology in the Feed Industry*. Proceedings of Alltech's 16th Annual Symposium, pp. 347-56. Nottingham University Press: Loughborough, UK.

Turman, E. J. and Andrews, F. N. (1955). Some effects of purified anterior pituitary growth hormone on swine. *J. Anim. Sci.* 14, 7-18.

Underwood, E. J. (1977). Chromium. In Underwood E. J. (Ed.), *Trace Elements in Human and Animal Nutrition*. Fourth edition, pp. 258-70. Academic Press: New York, NY, USA.

van Barneveld, R., Nuttall, J., Flinn, PC. and Osborne, B. (1998). NIR reflectance measurement of the digestible energy content of cereals for growing pigs. *J. Near Infrared Spec.* 7, 1-7 doi: (10.1255/jnirs.228).

van Barneveld, R. J. (1999). Physical and chemical characteristics of grains related to variability in energy and amino acid availability in pigs: a review. *Aust. J. Agr. Res.* 50, 667-87.

van Barneveld, R. J. and Ru, Y. J. (2003). Measurement of total and reactive lysine content of canola meal using near infrared spectroscopy. In Paterson J. E. (Ed.), *Manipulating Pig Production*. Volume IX, p. 157. Australasian Pig Science Association: Werribee, VIC, Australia.

van Barneveld, R. J. (2010). Mineral supplementation in modern pork production systems for optimum health and efficiency – the need for renewed focus. Proceedings of FeetFirst Sow Lameness Symposium II, pp. 66-81. Minneapolis, MN, USA.

van Barneveld, R. J., Graham, H. and Diffey, S. (2017). Predicting the nutritional quality of feed ingredients for pigs using near infra-red spectroscopy (NIRS) and chemical analysis. *Anim. Prod. Sci.* (In press).

van Barneveld, R. J., Nuttall, J. D., Flinn, P. C. and Osborne, B. G. (1999a). Near infrared reflectance measurement of the digestible energy content of cereals for growing pigs. *J. Near Infrared Spec.* 7, 1-7.

van Barneveld, R. J., Ru, Y. J., Szarvas, S. R. and Wyatt, G. F. (1999b). Effect of oil extraction process on the true ileal digestible reactive lysine content of canola meal. In Cranwell P. D. (Ed.), *Manipulating Pig Production*. Volume VII, p. 41. Australasian Pig Science Association: Werribee, VIC, Australia.

van Dijk, A. J., Everts, B., Nabuurs, M. J. A., Margry, R. J. C. F. and Beynen, A. C. (2001). Growth performance of weanling pigs fed spray-dried animal plasma: a review. *Livest. Prod. Sci.* 68, 263-74.

Wang, Y. Z., Xu, Z. R., Lin, W. X., Huang, H. Q. and Wang, Z. Q. (2004). Developmental gene expression of antimicrobial peptide PR-39 and effect of zinc oxide on gene regulation of PR-39 in piglets. *Asian-Australas. J. Anim. Sci.* 17, 1635-40.

Weldon, W. C., Lewis, A. J., Louis, G. F., Kovar, J. L., Giesemann, M. A. and Miller, P. S. (1994). Postpartum hypophagia in primiparous sows: I. Effects of gestation feeding level on feed intake, feeding behaviour, and plasma metabolite concentrations during lactation. *J. Anim. Sci.* 72, 387-194.

Westendorf, M. L. (2000). Food waste as swine feed. In Westendorf, M. L. (Ed.), *Food Waste to Animal Feed*, pp. 69-89. Iowa State University Press: Ames, IA, USA.

Williams, N. H., Patterson, J. and Foxcroft, G. (2005). Non-negotiables of gilt development. *Adv. Pork Prod.* 16, 281-9.

Wolter, B. F., Ellis, M., Corrigan, B. P. and DeDecker, J. M. (2002). The effect of birth weight and feeding of supplemental milk replacer to piglets during lactation on preweaning and postweaning growth performance and carcass characteristics. *J. Anim. Sci.* 80, 301-8.

Woods, A. L., Armstrong, T. A., Anderson, D. B., Elam, T. E. and Sutton, A. L. (2011). Case Study: Environmental benefits of ractopamine use in United States finisher swine. *Prof. Anim. Sci.* 27, 492-9.

Wu, G., Bazer, F. W., Burghardt, R. C., Johnson, G. A., Kim, S. W., Li, X. L., Satterfield, M. C. and Spencer, T. E. (2010). Impacts of amino acid on pregnancy outcome in pigs: mechanisms and implications for swine production. *J. Anim. Sci.* 88, E195-E204.

Wu, G., Bazer, F. W., Davis, T. A., Jaegar, L. A., Johnson, G. A., Kim, S. W., Knabe, D. A., Meininger, C. J., Spencer, T. E. and Yin, Y.-L. (2007). Important roles for the arginine family of amino acids in swine nutrition and production. *Livest. Sci.* 112, 8-22.

Wu, G., Bazer, F. W., Johnson, G. A., Knabe, D. A., Burghardt, R. C., Spencer, T. E., Li, X. L. and Wang, J. J. (2011). Important roles for l-glutamine in swine nutrition and production. *J. Anim. Sci.* 89, 2017-30.

Wu, G., Ott, T. L., Kanbe, D. A. and Bazer, F. W. (1999). Animal acid composition of the fetal
    pig. *J. Nutr.* 129, 1031–8.
Ytrehus, B., Carlson, C. S. and Ekman, S. (2007). Etiology and pathogenesis of
    osteochondrosis. *Vet. Pathol.* 44, 429–48.
Ytrehus, B., Grindflek, E., Teige, J., Stubsjøen, E., Grøndalen, T., Carlson, C. S. and Ekman,
    S. (2004). The effect of parentage on the prevalence, severity and location of lesions
    of osteochondrosis in swine. *J. Vet. Med. A Physiol. Pathol. Clin. Med.* 51, 188–95.

Chapter 2

Managing feed to optimize pig health

Sam Millet, Flanders Research Institute for Agriculture, Fisheries and Food (ILVO), Belgium; and Nadia Everaert, TERRA Teaching and Research Centre, Gembloux Agro-Bio Tech, Liège University, Belgium

1 Introduction

Feed is the most important cost factor in pig production. Optimal feeding goes far beyond merely providing the right set of nutrients to optimize the animal's growth: feeding plays a role in gastrointestinal diseases, such as weaning diarrhea and swine dysentery, but it also affects body condition, lameness, and immunity. Proper nutrition may increase resilience and prevent disease. But what comprises proper nutrition? There is no simple answer, as it often depends on the farm situation. While optimal performance has economic and ecological benefits and should be the main goal on farms with optimal health conditions, most farms need to balance trade-offs between feeding for optimal performance and minimizing the risk of health problems.

Feed management means choosing the right set of ingredients to meet the pigs' nutrient requirements. Defining those requirements may depend not only on the physiological status of the pig but also on its health status. At certain time points, pigs may require feeding below their requirements to improve their health. This chapter begins with a discussion of how balanced energy and protein levels are required for optimal health and performance (Section 2). A sufficient quantity and good quality of drinking water are also important in

http://dx.doi.org/10.19103/AS.2022.0103.11

this respect, but it is not discussed in this chapter. Vitamins and minerals are essential for physiological functions and metabolism, but the animal's health status may be affected when too much or too little of these micronutrients are provided relative to the animal's physiological requirements (Section 3). Fiber should not only be seen as a bulking agent but may also be considered in the context of intestinal health (Section 4). The choice of ingredients for minimal feed cost may differ from the choice for optimal health; certain ingredients are discussed in the health context of piglets (Section 5). In addition, the form and structure of the feed are important characteristics as they can affect both health and performance (Section 6).

2 Healthy energy and protein intake

Energy and amino acid levels are major determinants of pig diets, especially in relation to optimal performance. It is generally assumed that pigs eat to meet their energy needs. When energy level decreases, feed intake (FI) is likely to increase (Henry, 1985). For this reason, the requirements of other nutrients are generally expressed relative to energy. A balanced amino acid:energy ratio leads to optimal muscle deposition and efficient growth, with the health status of the animal possibly affecting the amino acid pattern. If one or more amino acids are provided below the requirement, muscle deposition will decrease without any evident health risks. In growing-finishing pigs, the dietary amino acid:energy ratio mainly affects feed efficiency, with a lesser effect on health. In growing-finishing pigs under heat stress, increasing the dietary nutrient density and lowering the crude protein (CP) level (without affecting the ileal digestible amino acid level) may decrease heat production, as dietary protein leads to higher heat increment (the heat production following feed consumption) than fat or starch do (Noblet et al., 1994; Renaudeau et al., 2008).

In piglets, the dietary CP level is of main concern, but FI itself is also crucial to overcome post-weaning diarrhea (PWD). In sows, energy intake and body condition may be determinants of longevity.

The importance of gastric pH in piglets

A low gastric pH serves as a barrier to pathogens attempting to enter the small intestine (Hansen et al., 2007). Low gastric pH is important for protein digestion, as it is key for denaturizing dietary proteins as the first step in the digestive process. The transformation of pepsinogen into pepsin occurs most efficiently at low pH. Moreover, pepsin is most effective at pH 2.0-3.5 (Lawlor et al., 2005). In newly weaned piglets, HCl secretion is limited and only starts after consumption of solid feed (Cranwell, 1985). As a result, pathogens may better survive the stomach environment and protein digestion may be impaired, leading to excess levels of protein escaping the small intestine where it becomes a substrate for potentially harmful bacteria in the large intestine. The dietary acid-binding capacity of piglet feed should therefore be kept as low as possible, which can be achieved by careful ingredient selection (Lawlor et al., 2005). Weaned pigs may also benefit from whey or lactose in starter diets, as the ongoing lactic acid production lowers gastric pH.

2.1 The protein level in piglet diets

Prior to weaning and in the first weeks afterward, piglets eat little solid feed, which hampers their growth potential. To maximize performance, they need high dietary amino acid levels. However, if the protein level is too high, gastric functioning can be negatively affected and can result in an increased risk for intestinal problems. In piglets, high CP levels have been shown to increase the risk for PWD (Nyachoti et al., 2006; Wellock et al., 2007, 2008a; Heo et al., 2009; Rist et al., 2013). Different mechanisms may be at play here: first, protein increases the buffering capacity in the stomach, with higher pH leading to impaired protein digestion and greater passage of potential pathogens, leading to the growth of opportunistic pathogens like *E. coli* in the small intestine. The toxins produced by pathogenic *E. coli* cause elevated secretion of water and chloride from small intestinal crypt cells, leading to secretory diarrhea (Sun and Kim, 2017). Second, higher protein levels may lead to higher amounts of undigested protein entering the large intestine. Protein fermentation, accompanied by the formation of potentially harmful compounds, is regarded as a factor in the development of enteric diseases, especially under stressful conditions (Rist et al., 2013). Protein fermentation in the large intestine, with its potentially toxic metabolites, is reduced along with CP level (Sun and Kim, 2017). Lowering the CP level while maintaining essential amino acid levels is the first way to reduce the risk of diarrhea without affecting the performance (Htoo et al., 2007).

When reducing the protein level, good knowledge of essential amino acid requirements is essential for maintaining performance. Even then, protein reduction without affecting performance is only possible within limits: Jansman et al. (2016) could not reduce CP level below 160 g/kg. Gloaguen et al. (2014) achieved a reduction in CP for piglets weighing 10-20 kg from 176 g/ kg to 135 g/ kg without significantly affecting performance, but in that trial, the lysine level was below requirement. At lower CP levels, semi-essential or even nonessential amino acids may become limiting (Wu, 2014). At some point, nitrogen itself may limit growth. Lysine makes up 7.6 g/100 g of protein in the carcass (Mahan and Shields, 1998). As a practical guideline, a maximal

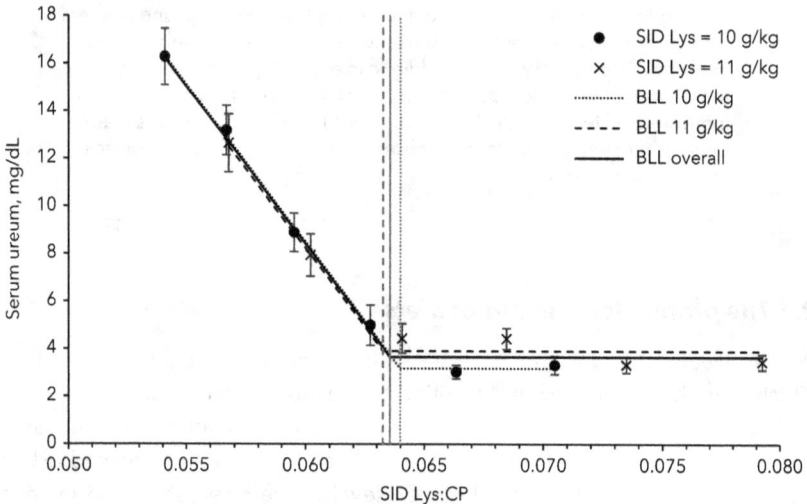

Figure 1 Effect of dietary standardized ileal digestible (SID) Lys to CP ratio on serum urea levels in piglets at 7 weeks of age. Piglets receiving 10 g SID Lys/kg or 11 g SID Lys/kg diet are presented, together with their corresponding and overall broken line models with linear ascending portions. Vertical gray lines represent breakpoints. Urea level decreases with decreasing CP level (=increasing SID Lys:CP ratio) until a minimum is reached. This indicates the point of maximal nitrogen retention and minimal protein breakdown. A logical consequence is that low CP diets for piglets may contain amino acid levels below the requirement for maximal growth and efficiency. In this period, in our opinion, it is a valid choice to feed below the requirements for optimal growth and as such decrease the risk for gastrointestinal disturbance. Source: Figure from Millet et al. (2018), with permission from Oxford University Press.

standardised ileal digestible (SID) lysine:CP ratio of 0.064 (SID Lys:ATTD CP = 0.079) can be used in piglet diets (Millet et al., 2018). At this point, serum urea level (as a measure of protein breakdown) is the lowest (Fig. 1). In our opinion, the optimal CP level in the diet may depend on the farm situation. In good health conditions, maximal performance can be achieved by using high dietary amino acid levels and possibly also higher CP levels. For most herds, however, nutritionists will lower the CP level to decrease the risk of diarrhea. This implies that suboptimal growth performance may be accepted to decrease the risk for disease and mortality. Furthermore, feed costs may be reduced by adapting the dietary SID lysine level to this lower CP level.

2.2 Amino acid levels for optimal health

In addition to protein biosynthesis, amino acids exert functions in regulating immunity, anti-oxidation, and redox regulation (Liu et al., 2017). According to Goodband et al. (2014), increase in amino acid requirements for immune function in times of disease are more than offset by the associated decrease in protein accretion and increased muscle protein degradation, which implies there would be no need for increased dietary amino acid levels. Still, the balance between amino acids may be altered. A study by van der Meer et al. (2016) suggested that methionine, threonine, and tryptophan supplementation improved the performance of pigs especially under poor sanitary conditions, which indicates again that the ideal amino acid pattern may depend on the health status of the animals. Similarly, supplementation with threonine, methionine, and tryptophan improved immune status in weaned pigs challenged with *Salmonella* (Rodrigues et al., 2021). The disadvantage of these studies is that it is impossible to separate the effect of a single amino acid. Particular amino acids may exert beneficial effects apart from their function as a building block for protein synthesis. Some reviews have been written on the role of individual amino acids in immune functioning and pig health (Li et al., 2007; Liu et al., 2017; Le Floc'h et al., 2018). Here, in the following sections, we describe three amino acids that have been shown to be important for maintaining health.

2.2.1 Glutamine

Glutamine, a unique amino acid, serves among others as a preferred respiratory fuel for cells that proliferate rapidly, like enterocytes and lymphocytes (Lacey and Wilmore, 1990; Wu et al., 1996). Even though it is abundantly present in the body, during stress the individual's body may not be able to produce sufficient amounts of this amino acid. Glutamine may therefore become a 'conditionally essential' amino acid in critically ill animals and during periods of stress such as the weaning period (Lacey and Wilmore, 1990). Sufficient glutamine intake may be needed for intestinal integrity. In research on piglets, glutamine supplementation beneficially affected the performance of early weaning piglets in the first 2 weeks after weaning, especially when fed a low protein diet (Millet et al., 2020a). L-glutamine in creep feed and nursery diet was shown to improve feed conversion in the first 3 weeks post-weaning, possibly due to improved gut health (Cabrera et al., 2013). Johnson and Lay (2017) observed that intestinal morphology was improved by L-glutamine supplementation and therefore stated that L-glutamine in piglet diets can be a cost-effective alternative to antibiotics following weaning and transport. Recently Ji et al. (2019) reviewed the studies of glutamine in weaning pigs.

2.2.2 Threonine

Threonine is considered an important amino acid for intestinal health. During acute intestinal inflammation, threonine fluxes across the portal-drained viscera (PDV) increase. Threonine is needed for the synthesis of mucin and plasma immunoglobulin production in animals (Li et al., 2007; Trevisi et al., 2015). Hamard et al. (2010) found that moderate threonine deficiency affected the functionality of the small intestine of piglets. Similarly, dietary L-threonine supplementation attenuated inflammatory responses, facilitated Muc2 synthesis, and promoted goblet cell differentiation in the ileum of intrauterine-growth-retarded weanling piglets (Zhang et al., 2019). Furthermore, Trevisi et al. (2015) showed an altered microbiota composition as evidenced by a reduced *E. coli* shedding in pigs fed at higher threonine levels.

2.2.3 Arginine

Arginine plays an important role in intestinal health: it protects against inflammatory stress and inflammation, modulates mTOR in intestinal tissue and the intestinal inflammatory response, and attenuates villus atrophy (Le Floc'h et al., 2018). Part of its effect is due to its role in the formation of nitric oxide, which acts as a vasodilator, signaling molecule, and mediator of immune response (Le Floc'h et al., 2018). It is considered a conditionally essential amino acid (NRC, 2012). Liu et al. (2008) observed that arginine supplemented in the piglet diet attenuated the negative effects of *E. coli* lipopolysaccharide on the intestinal inflammatory response and intestinal morphology. Similarly, Zheng et al. (2017) observed a positive effect in an experimental model of oxidative stress in piglets by decreasing inflammatory cytokine expression.

Recently, additional free amino acids have become commercially available for use in feeds. This allows for more targeted amino acid nutrition. This evolution is expected to lead to more research into their functions beyond muscle formation.

2.3 Feed intake as a driver for healthy weaning

The stressful period shortly after weaning is characterized by reduced FI. Weaning age, weaning stress, FI, and diet composition all affect intestinal integrity after weaning (Wijtten et al., 2011). FI affects the flow of nutrients through the intestinal tract and intestinal morphology. Ensuring sufficient nutrient intake after weaning is a challenge, but it remains key to a proper gut structure and function and to the prevention of several problems associated with weaning. This process starts before weaning. Bruininx et al. (2002) observed that piglets that ate creep feed before weaning showed more favorable FI behavior and thus a higher daily gain in the first 8 days post-weaning than the non-eaters. Several management strategies in the pre- and post-weaning phases have been investigated to ensure sufficient nutrient intake after weaning. For instance, restricted feeder space increases competition and reduces FI, and thus negatively influences gut development and health as well as performance

(He et al., 2016). Other factors include creep feed provision, the composition of creep feed and weaning diet, mat feeding, sow management, barn climate, pellet size, nursery placement strategies, water access points and location, and diet form. Flavor imprinting is another strategy that may stimulate FI of weaned piglets and reduce diarrhea incidence (Oostindjer et al., 2010). An overview of these strategies was published recently by Wensley et al. (2021a,b).

The amino acid profile may also have a profound effect on FI. Especially when the ratio between amino acids is unbalanced, the animals react by decreasing their FI. Especially excess leucine depresses FI and the growth of piglets (Goodband et al., 2019). High brain leucine levels reduce FI by hypothalamic mTOR signaling (Cota et al., 2006). This negative effect can be (partially) counteracted by adding more valine to the diet (Millet et al., 2015). Valine has been shown to increase FI (Gloaguen et al., 2011; Millet et al., 2020b). It is important to note that the ratios between amino acids may be more important than their absolute values. In recent research, FI was largely affected by decreasing lysine levels when piglets were fed a high protein diet and thus an excessive leucine:lysine ratio, while this effect was less pronounced when they were fed a diet with a lower CP level and more balanced amino acid profile (Millet et al., 2020b) (Fig. 2). In addition, the sanitary conditions may affect the

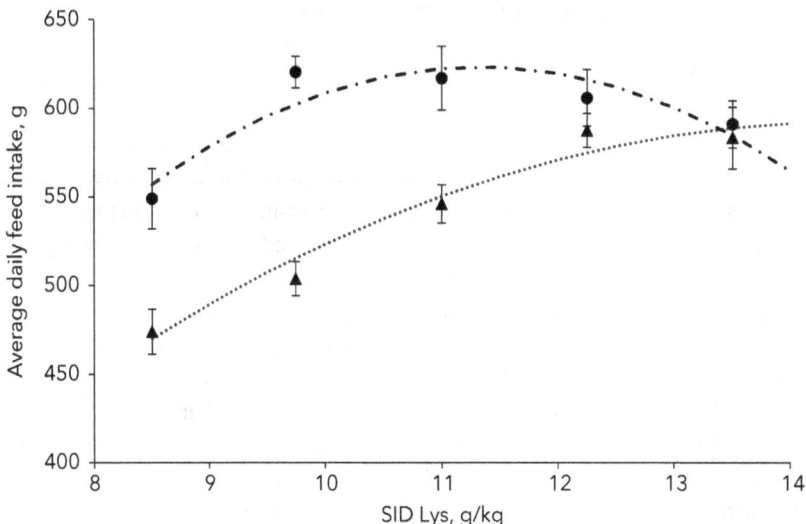

Figure 2 Measured means (▲ for Exp. 1, ● for Exp. 2) and standard error bars for each treatment are presented, with their fitted quadratic polynomial models. With decreasing lysine levels, feed intake decreased especially in experiment 1. In this experiment, a high-protein diet was used and with decreasing lysine levels the amino acid ratios were less and less balanced. In experiment 2, the balance between amino acids was maintained better by using a low CP diet and free amino acids. Source: Reprinted from Millet et al. (2020b), with permission from Elsevier.

optimal AA ratio. For example, a higher optimal ratio of SID threonine:lysine for weaned piglets was shown under unclean sanitary conditions compared to clean sanitary conditions. Poor air quality may have affected FI and growth performance (Jayaraman et al., 2015).

2.4 Feed and energy intake and body condition of sows

In sows, optimal body condition is necessary to maintain health and longevity. While sufficient amino acid provision is important, daily energy intake may be even more important: 'thin sow syndrome,' 'fat sow syndrome,' and 'second parity syndrome' have all been related to the dynamics of body condition management (Charette et al., 1996). Body condition is managed by controlling FI and the energy level in the diet. During gestation, energy and amino acids are needed to grow the fetuses and for the sow's own growth, but a good diet is also needed to build sufficient body reserves to compensate for a possible nutritional deficit that may occur in the following lactation (Dourmad et al., 1994). During lactation, sows mobilize fat and muscle mass to compensate for the negative energy balance, as FI during lactation is often not high enough to sustain the milk production needed for large litters (Eissen et al., 2000; Bergsma et al., 2009). Body condition at the end of lactation thus determines the energy requirements during gestation. The speed of body condition recovery is still a matter of research. In general, studies suggest it may be good to increase the FI shortly after weaning (Leal et al., 2019), but others claim this may negatively affect embryo survival and litter size (Mallmann et al., 2020). The importance of increasing the total FI of sows during late gestation has been questioned by Goodband et al. (2013). They suggest increasing the amino acid intake, without the sows becoming too fat, as overfeeding in gestation often leads to lower FI during lactation (Goodband et al., 2013). A logical conclusion is that two-phase feeding during gestation may be optimal.

Sows that are too thin at the start of lactation have been associated with a higher number of stillborn piglets (Maes et al., 2004b), lower colostrum production (Decaluwe et al., 2014), and more shoulder lesions (Bonde et al., 2004; Zurbrigg, 2006; KilBride et al., 2009; Rioja-Lang et al., 2018). In contrast, fat sows or sows overfed during gestation have lower FI and less milk production during lactation (Weldon et al., 1994). Moreover, excessive fattening during late pregnancy is related to a longer farrowing duration (Oliviero et al., 2010), with a higher risk for perinatal mortality. Decaluwé et al. (2013) observed that colostrum yield was negatively associated with the use of body protein before farrowing, whereas it was positively associated with the use of body protein on the first day of lactation. The restriction of feeding prior to parturition significantly reduces the risk of postpartum dysgalactia syndrome (PDS) (Peltoniemi et al., 2007). In conclusion, sows need optimal body condition management during gestation and lactation. Both overfeeding and underfeeding pose risks for sow health and longevity. For further reading on this topic, we suggest the review paper of Solà-Oriol and Gasa (2017).

3 Mineral and vitamin nutrition for resilient animals

Minerals and vitamins both play an important role in animal metabolism. Deficiencies as well as toxicities are problematic and lead to different types of symptoms. In contemporary pig production, such symptoms are rare, as premixes meant to supplement the diet contain a balanced mixture of vitamins and minerals. Therefore, we focus only on a few important macro- and micro minerals that can still be linked to a higher risk for production diseases. For advised levels, we refer to NRC (2012), CVB (2020), GfE (German Society for Nutrition Physiology, 2006), and local recommendations.

3.1 Calcium and phosphorus

Calcium (Ca) and phosphorus (P) are essential minerals for optimal bone mineralization. Both minerals have to be considered together when looking at the requirements as they are homeostatically regulated. The Ca:P ratio should remain constant, with a minimum of 1:1 (van Riet et al., 2013). Since phosphorus excreted in the manure is considered a pollutant, the use of low phosphorus diets in growing-finishing pigs is encouraged. The use of phytase can be particularly helpful for this. More than 95% of body calcium is stored in the bones, while phosphorus is mainly stored in the bones as well as in lean tissue and organs (Bikker and Blok, 2017). Excessive calcium, in comparison to phosphorus, is excreted through the urine. An unbalanced calcium:phosphorus ratio may lead to urolithiasis (Lorenzett et al., 2019). Uroliths may be present in a larger proportion of fattening pigs than usually expected (Vrielinck et al., 2019), despite the small number of reports of obstructive urolithiasis in pigs (Maes et al., 2004a). The mineral composition of the diet, together with the amount of drinking water and the urinary pH, is considered the major risk factors for urolithiasis (Maes et al., 2004a). When pigs eat diets deficient in phosphorus, priority is given to phosphorus retention in soft tissue instead of phosphorus retention in bones (Misiura et al., 2020). Insufficient calcium and phosphorus levels thus lead to poor bone mineralization and eventually osteomalacia or hind leg paralysis (NRC, 2012). The exact role of calcium or phosphorus shortages is not entirely clear, as the reason for lameness is mostly multifactorial (Wegner et al., 2020). Apart from its impacts on animal welfare, lameness is an important reason for the early culling of sows (Pluym et al., 2013).

Recommended levels for growing pigs are based on a total phosphorus level of 5.3 g/kg empty body weight gain (Bikker and Blok, 2017). Translating this to dietary levels may not be easy. Especially in highly efficient animals, the ingested calcium and phosphorus level may be lower than optimal because of their low FI in comparison to their growth rate. The advice is therefore to compensate for the lower FI per kilogram gain in highly efficient animals through increased calcium and phosphorus levels per kilogram of feed (Bikker and Blok, 2017). In sows, mineral requirements vary throughout a sows' cycle, with the highest requirements at the end of gestation and during lactation. The bone acts as a reservoir with bone mineralization during gestation and mobilization in lactation, as reflected in the concentration of bone mineralization markers (van Riet et al., 2016b). Because bone mineralization is dynamic, assessment of 'optimal' mineralization is difficult.

Feeding below calcium and phosphorus requirements has been advised for young piglets, as calcium sources such as limestone are characterized by a high buffering capacity. Jongbloed et al. (2003) recommended the use of a maximum calcium content below the optimum, while accepting a transient reduction in bone mineralization. This was adopted in feeding recommendations of Bikker and Blok: instead of the calculated requirements of 5.9 g/kg and 3.4 g/kg for calcium and phosphorus, respectively, they suggest using 5.0 g/kg and 3.0 g/kg during the first weeks after weaning to reduce the calcium content and buffer capacity of the weaning diet (Bikker and Blok, 2017).

3.2 Zinc and copper

The microminerals zinc and copper have received considerable attention in recent years. They are involved in many metabolic functions as cofactors for a variety of enzymes (Jondreville et al., 2003) and play a crucial role in bone formation and claw integrity (van Riet et al., 2013, 2016a), skin development (Tucker and Salmon, 1955), and immune function (Kloubert et al., 2018). To meet the copper requirements, Bikker and Jongbloed (2013) calculated copper levels of 18 mg/kg for lactating sows and 12 mg/kg for other pigs, while maximum allowed levels in the EU are 150 mg/kg for piglets during the first 4 weeks after weaning, 100 mg/kg until 8 weeks after weaning, and 25 mg/kg for other animals (European Commission, 2018). For zinc, the recommended dietary levels to meet the requirements vary from 65 mg/kg in growing-finishing pigs to 85 mg/kg in lactating sows, with maximum allowed levels of 150 mg/kg for piglets and sows and 120 mg/kg in other animals.

However, most of these ingested microminerals are excreted through the feces. In the soil, they may be toxic to plants and microorganisms and have thus become an environmental concern (Jondreville et al., 2003). Dietary levels are therefore restricted by legislation.

Although the use of microminerals as growth promoters has been criticized, especially zinc oxide (ZnO) at 'pharmacological' concentrations (2,500 mg/kg) has been shown to be effective in fighting PWD. The mechanism of action is not well understood (Debski, 2016), but it seems to be multifactorial and goes beyond the fulfilment of daily nutritional requirements (Bonetti et al., 2021). It has been shown that ZnO exerts anti-inflammatory effects (Lallès and Montoya, 2021) and improves intestinal morphology and intestinal barrier function (Hu et al., 2013; Song et al., 2015), thereby probably preventing translocation of bacteria through the intestinal barrier and increasing the activity of digestive enzymes (Hedemann et al., 2006). While ZnO does have antibacterial properties, its ability to reduce *E. coli*, even at high doses, is questioned (Liedtke and Vahjen, 2012). Apart from inorganic forms, such as zinc sulfate and zinc chloride, organic microminerals, such as Zn-methionine or Zn-lysine, are now commonly available in commercial pig feeding. They have been claimed to be more effective than inorganic mineral sources in improving intestinal barrier function (Pearce et al., 2015), immune function (Richards et al., 2010), or claw health (Lisgara et al., 2016), although many studies showed little or no difference in the bioavailability of organic versus inorganic mineral sources (Richards et al., 2010; Schlegel et al., 2013). Similarly, several researchers did not see differences between zinc source on immune function parameters (van Heugten et al., 2003) or performance (Hollis et al., 2005; Holen et al., 2018).

The common usage of ZnO at pharmacological levels in pig farming has raised several concerns. Apart from the resistance of microorganisms to ZnO itself, the use of ZnO in animal diets seems to promote the spread of antibiotic resistance in *E. coli* (Bednorz et al., 2013). Together with environmental concerns due to the excretion of excess zinc (Bonetti et al., 2021), it makes ZnO a less sustainable alternative for antibiotics. This has led to a phase-out of therapeutic ZnO use in Europe by June 2022. To cope with reduced copper and zinc levels in European legislation, new forms such as nanoparticles have been developed and marketed. Positive effects on performance and health parameters have been reported (Wang et al., 2018; Bai et al., 2019; Pei et al., 2019), although most of these studies were carried out with concentrations higher than 150 mg/kg. In a study by Milani et al. (2017), doses up to 60 mg/kg were not able to improve diarrhea scores (in contrast to therapeutic levels of ZnO). In summary, there is at this moment no sustainable mineral-based alternative for the pharmacological use of ZnO, although research on alternative forms and doses is still ongoing.

3.3 Optimal vitamin levels

The fat-soluble vitamins (A, D, E, K) can be accumulated in the body (mainly the liver), with a possible risk of toxicity, while this is not the case for the water-soluble vitamins (all Bs and C). Although little is known regarding the quantitative need for vitamins for supporting the optimal development of innate and adaptive immune responses, the use of a premix rich in vitamins has largely eliminated the risk of vitamin deficiency in modern intensive pig production (Lauridsen et al., 2021). However, the rapid genetic improvement and high performance of modern breeds may lead to disrupted redox balance and uncontrolled inflammation. This increases the need to control reactive oxygen species and peroxides. Antioxidant vitamins (vitamin E and vitamin C) may therefore be considered as a strategy for enhanced enteric immunity and inflammation control (Lauridsen, 2019). Critical periods are soon after birth, especially under limited colostrum intake, and post-weaning under conditions of limited FI and impaired digestive functioning. For larger pigs, heat stress increases micronutrient needs due to increased utilization coupled with reduced FI (Cottrell et al., 2015). This may trigger oxidative stress, including in the gut (Liu et al., 2016). Besides the abovementioned vitamins, selenium is also a well-known biological antioxidant. Results of supplementing sows with either vitamin E or selenium have included a higher number of piglets born and weaned, increased piglet weight at weaning, enhanced milk fat content, better humoral immune function (IgG and IgA), and improved oxidant activity of sows and piglets (Mavromatis et al., 1999; Pinelli-Saavedra et al., 2008; Wang et al., 2017). Selenium and vitamin E supplementation in growing pigs has been shown to reduce intestinal leakiness caused by heat stress (Liu et al., 2016). Vitamins A and D can also regulate intestinal immune functions (Lauridsen et al., 2021). Indeed, in a weanling pig model, supplementation with vitamin D_3 or its metabolite 25OHD$_3$ increased leukocyte cell numbers and had a potential impact on systemic and mucosal antimicrobial responses (Konowalchuk et al., 2013). For an extensive review of the role of vitamins on gastrointestinal functionality and health of pigs, we refer to the review of Lauridsen et al. (2021). Optimal levels of each of the vitamins are discussed by Isabel et al. (2012).

4 Fiber, an overlooked nutrient?

Dietary fiber (DF) is defined as 'carbohydrate polymers and lignin that are not hydrolysed by endogenous enzymes and are substrates for bacterial fermentation in the hindgut' (Lupton et al., 2009). It includes plant cell walls, as well as resistant starch and non-digestible oligosaccharides, which include plant cell contents and include fructo-oligosaccharides (Agyekum and Nyachoti, 2017). The main cell wall polysaccharides in cereal grains are cellulose, arabinoxylans, and mixed linked β-(1,3) (1,4)-D-glucans, while cell walls of oilseeds, oilseed meal, and leguminous crops contain cellulose, pectic polysaccharides, lignin, and xyloglucans. Legumes also contain significant quantities of galacto-oligosaccharides including raffinose, stachyose, and verbascose (Navarro et al., 2019).

DF is classified based on physicochemical properties, that is viscosity, hydration, and fermentability, in order to provide more information on its metabolic and physiological activities (Agyekum and Nyachoti, 2017). Solubility is a major factor: non-starch polysaccharides (NSP) can be classified as insoluble (cellulose and some hemicelluloses) or soluble (pectins, gums, and β-glucans) based on their solubility in water or weak alkali (Bach Knudsen, 2001).

Contradictory results on the use of DF have been reported, which can be explained by the influence of different factors such as variation in ingredient composition, the inclusion of different fiber types with different physicochemical properties, inclusion levels, and the matrix provided by the other ingredients. The review of Agyekum and Nyachoti (2017) provides more information on the physicochemical properties of fibers and the mode of action on digestive physiology, health, and performance of pigs. For an extensive review of the structure and characteristics of the carbohydrates that compose DF, see Navarro et al. (2019).

4.1 Role of fibers in the digestive tract

Soluble fibers may increase digesta viscosity, thus hampering nutrient digestion and absorption due to increased cell exfoliation causing villus atrophy (Jha and Berrocoso, 2016). In addition, they may hinder contact between digestive enzymes and nutrients. In general, soluble fiber appears to have a greater impact on starch digestibility than insoluble fiber due to the increase in luminal viscosity and the nutrient-encapsulation effect of intact cell wall polysaccharides (Agyekum and Nyachoti, 2017). Insoluble fiber increases digesta passage rate, allowing less time for mixing of digestive enzymes and dietary components, which also leads to decreased nutrient digestibility. Fiber may reduce apparent ileal protein digestibility through increased endogenous and exogenous nitrogen and amino acid losses. In the large intestine, fiber may stimulate bacterial protein synthesis, leading to decreased apparent total tract protein digestibility. Inclusion of fibrous feedstuffs in the diet, therefore, shifts nitrogen excretion from urea in urine to bacterial protein in feces (Canh et al., 1997). Furthermore, due to the increased workload of the secretary organs, DF may increase hypertrophy, which increases visceral organ mass and cell exfoliation (Agyekum and Nyachoti, 2017). However, DF can contribute to some extent to the energy supply of the pigs due to fermentation by the microbiota in the distal small intestine and large intestine. This produces short-chain fatty acids (SCFA; mainly acetate, propionate, and butyrate) although energy losses also occur due to the production of methane and hydrogen, as well as heat from the fermentation process. While butyrate is used as an energy source for colonic cell proliferation, propionate and a certain amount of butyrate are both used for hepatic gluconeogenesis and nearly two-thirds of acetate is metabolized in muscle cells (Slavin, 2013).

Another physiological effect of fibers is their effect on FI and satiety. FI is typically influenced by dietary energy content (Henry, 1985), but the compensation for reduced energy levels is often not complete. A prolonged satiety effect has been observed by including DF (Danielsen and Vestergaard, 2001), which has been ascribed to their ability to delay gastric emptying, the increased swelling of the stomach content, and the release of satiety-related gut hormones upon the increased abundance of SCFA in the hindgut (Agyekum and Nyachoti, 2017). Therefore, DF is often incorporated at higher levels in the diet of gestating sows to limit a high energy intake, to reduce the sensation of hunger, and to overcome the aggression and behavioral problems that are associated with restricted feeding (de Leeuw et al., 2008).

4.2 Fibers and gut health

The microbial fermentation by bacteria, fungi, protozoa, and archaea results in the production of SCFA, branched-chain fatty acids (BCFA), lactate, amines, indoles, phenols, and various gases like hydrogen, carbon dioxide, and methane (Jha and Berrocoso, 2016). Soluble fiber generally is more extensively and more quickly fermented than insoluble fiber. Fiber inclusion modulates the gut microbiota. For example, resistant starch selectively promotes bifidobacteria and lactobacilli in weaned and growing pigs (Metzler-Zebeli et al., 2019). The fiber fermentation and subsequent production of SCFA contribute to the maintenance of an anaerobic environment, preventing the proliferation of facultative anaerobic pathogens such as *E. coli* or *Salmonella* (Jha et al., 2019). Moreover, the microbial metabolites can promote the proliferation of the mucosal epithelium by stimulating goblet cell proliferation, increasing villus height and hence the absorptive surface area, and in some cases decreased intestinal permeability. Indeed, supplementation with insoluble fiber sources improves gut morphology, with a rise of the villus height to crypt depth ratio as well as an increased number of colonic goblet cells and mucin areas (Hedemann et al., 2006; Chen et al., 2013). The SCFA are also potential immune mediators for the host due to their immunomodulatory and anti-inflammatory properties (Correa-Oliveira et al., 2016). For example, the inclusion of oat bran in the diet of growing pigs modulated the colonic microbiota and decreased mRNA expression of IL-8 in the caecum and reduced IL-8, NF-κB, and TNF-α gene levels in the colon, ameliorating inflammatory responses in the hindgut of growing pigs (He et al., 2018). However, the understanding of how microbiota improve immune function is in its infancy (Jha et al., 2019), especially in pigs, as most knowledge comes from rodent studies as a model for humans.

It is important to note that not all studies support the ability of DF to promote intestinal health (Agyekum and Nyachoti, 2017). Conflicting results have been observed on the inclusion of soluble fiber and *E. coli* proliferation (Hopwood et al., 2004) and occurrence of PWD (Wellock et al., 2008b; Halas et al., 2009). Moreover, the inclusion of resistant starch increased the pro-inflammatory cytokine IL-1β, indicating a negative impact on the immune system (Sun et al., 2015). Pectin was associated with a lower number of goblet cells (Święch et al., 2012) and consequently a reduction of mucin areas in the small intestine crypts (Hedemann et al., 2006).

4.3 Optimizing the use of fibers

To improve the nutritive value of high-fiber diets, several strategies can be applied such as the formulation of diets based on net energy or available nutrient basis, particle size reduction, and pelleting. However, these may counteract the health effects (see below). Several enzyme preparations are commercially available to degrade not only the fiber in pig feeds, such as xylanases, β-glucanases, but also combinations with β-mannanases, α-galactosidases, and pectinase. Also in this field – contrary to poultry studies – inconsistent results have been reported on nutrient utilization and growth performance (Agyekum and Nyachoti, 2017). We hypothesize that one of the reasons may be that these enzymes only speed up digestion by digesting components in the small intestine that would have normally broken down in the large intestine through fermentation and bacterial enzymes.

5 Ingredient composition

The pig diet is comprised of a mix of ingredients. Especially in piglet diets, the choice of ingredients in relation to weaning issues has been studied, although it is often difficult to disconnect the effect of the nutrients it brings from other properties. For example, cooked white rice has consistently been shown to protect against weaning diarrhea (McDonald et al., 1999; Hopwood et al., 2004; Montagne et al., 2004) and *Brachyspira hyodysenteriae* infection (Pluske et al., 1996; Siba et al., 1996). Still, it is difficult to conclude whether this is the result of the low fiber content of white rice or other factors inherent in this ingredient.

In particular, the search for optimal protein sources is ongoing. Animal-based products are of value, such as dairy products, spray-dried plasma (SDP), and egg powder. Spray-dried plasma, a protein-rich product obtained from the industrial fractionation of blood from healthy animals, has been shown to have positive effects on the health and performance of piglets (reviewed by Pérez-Bosque et al. [2016]). Skim milk powder improved intestinal morphology compared to hydrolyzed feather meal (Vente-Spreeuwenberg et al., 2004).

Immunized egg powder has been shown to protect against *E. coli* diarrhea (Yokoyama et al., 1992; Marquardt et al., 1999; Wang et al., 2019) although not in all studies (Aluko et al., 2017).

In comparison with animal protein sources, plant protein sources have higher levels of DF and might contain anti-nutritional factors (Montagne et al., 2004). Soybean meal – although easily digested and an interesting ingredient in growing pig diets – has been linked to PWD and its use in piglet diets is therefore limited (Friesen et al., 1993). The link between soy and PWD may be an aberrant immune response to soy proteins, leading to hypersensitivity (Engle, 1994). Moreover, the use of soybean products in pig diets is criticized because of its environmental impact. The soybean meal used in Europe is predominantly produced overseas. Especially when soy production leads to deforestation, it may be a threat to tropical biodiversity and may increase the global warming potential, energy, and land use of the pig feed (Fearnside, 2002; van Zanten et al., 2015). Ingredients that have shown potential to replace soybean meal include fermented soybean meal (Kiers et al., 2003; Feng et al., 2007; Xie et al., 2017; Wang et al., 2020), (hydrolyzed) wheat gluten (Blasco et al., 2005; Wang et al., 2011), and potato protein (Pedersen and Lindberg, 2004; de Souza et al., 2019). Recently, Cruz et al. (2019) showed that also yeast, even up to 40% of dietary CP, may be a valuable protein source that positively influences intestinal morphology and fecal consistency. It should be taken into account that distribution of these specialty ingredients has commercial value and research bias can therefore be anticipated (Resnik, 2000). Apart from the major ingredients, a great deal of research has been performed on the development and evaluation of feed additives in pig production. Especially for piglets, the search for alternatives to antibiotics is ongoing. They include pre-, pro-, or synbiotics, organic acids, exogenous enzymes, essential oils, and phytochemicals. The mechanisms associated with the beneficial effects of pre-, pro-, and synbiotics include manipulation of intestinal microbial communities (Guevarra et al., 2019), the neutralization of toxins, restriction of colonization and adhesion of opportunistic pathogens (Barba-Vidal et al., 2018), enhancement of the mucosal barrier (Trevisi et al., 2017), and stimulation of the immune system (Lessard et al., 2009). The mode of action of plant-derived substances, that is essential oils, organic acids, or phytochemicals, is mostly attributed to their antimicrobial and antioxidant properties, with the former stabilizing the intestinal microbiota (Lillehoj et al., 2018). In addition, organic acids may lower both the dietary and the digesta pH and promote intestinal morphology in weaned pigs (Tugnoli et al., 2020). Feed enzymes are also considered to play a positive role in gut health in pigs by improving nutrient digestibility, producing possible prebiotic oligosaccharides, or modulating gut microbiota, as reviewed by Kiarie et al. (2013). Pre-, pro-, and synbiotics in animal nutrition have been reviewed by Markowiak and Slizewska (2018). Although several positive effects of these supplements have been observed, not every study confirms this, as for example shown in a probiotic study that failed to reduce *Salmonella* Typhimurium prevalence (Peeters et al., 2019). Essential oils, organic acids, and phytochemicals were reviewed by Lillehoj et al. (2018), Omonijo et al. (2018), and Tugnoli et al. (2020).

6 Case study: the importance of feed structure for health and performance

Throughout this chapter, we have shown that optimal nutrition may depend on the situation. Under stressful conditions such as the weaning period, diets may be formulated either with more expensive ingredients or below the nutrient requirements for optimal growth performance. The search for the optimal diet form and feed structure may affect both performance and gastrointestinal health, as was observed in trials with finishing pigs at Flanders Research Institute for Agriculture, Fisheries and Food (ILVO) and Ghent University. In our first trial (Millet et al., 2012b), we evaluated the effect of particle size distribution and fiber content on gastric mucosa integrity. To do this, we formulated a high- and low-fiber diet with similar net energy and amino acid levels. The diets were ground with a hammer mill provided with a 1.5 mm or 6 mm screen. This resulted in four treatment groups: low-fiber finely ground, low-fiber coarsely ground, high-fiber finely ground, and high-fiber coarsely ground. Only one diet, the coarsely ground high-fiber diet, was successful in limiting the number of gastric lesions (Table 1). Unfortunately, the feed efficiency, expressed as kilogram of feed per kilogram of carcass gain, was significantly worse on this coarsely ground high-fiber diet compared to the other diets. This trial showed that feed structure, as a combination of physical and chemical characteristics, can affect gastric health. It also showed that a diet for optimal health does not always lead to optimal performance. In a similar trial (Millet et al., 2012a), we compared a mash diet with an expanded diet and a crumbled pelleted diet. Expanding and pelleting the diet led to better feed efficiency. However, these technological treatments had a negative impact on gastric mucosa integrity (worse gastric lesion scores) and were associated with increased *Helicobacter suis* colonization. Thus, in both trials, the diets with the best performance negatively affected gastric health. Both were probably the result of the dietary particle size, as pelleting and crumbling can reduce the particle size of the diet (Wolf et al., 2010; Vukmirović et al., 2017). Smaller particles lead to a higher surface area, allowing better contact with digestive enzymes (Vukmirović et al., 2017). Conversely, they also result in higher fluidity in the stomach, eventually leading to a less distinct pH gradient in the stomach and the formation of gastric ulcers in the esophageal part (Maxwell et al., 1970). Reduced particle size led to similar results with respect to *Salmonella* infections. Pigs fed a non-pelleted diet were less susceptible to *Salmonella enterica* serovar Typhimurium colonization,

Table 1 Frequency of macroscopic lesion scores in the stomach in pigs consuming diets with a low- or a high-fiber content that were finely or coarsely ground (n = 47 per treatment)

Fiber content	Low		High		P	
Particle size	Fine	Coarse	Fine	Coarse	Fibre content	Particle size
Lesion score 0	4.3	2.1	0.0	46.8	<0.001	<0.001
Lesion score 1	8.5	12.8	25.5	27.7		
Lesion score 2	55.3	53.2	59.6	19.1		
Lesion score 3	21.3	17.0	6.4	2.1		
Lesion score 4	4.3	8.5	0.0	2.1		
Lesion score 5	6.4	6.4	8.5	2.1		

Note: Only pig receiving the diet with high-fiber level that was coarsely ground showed better scores for the gastric mucosa.
Source: Adapted from: Millet et al. (2012b), with permission from Elsevier.

especially when feed was coarsely ground (Mikkelsen et al., 2004; Hedemann et al., 2005). Similarly, in a study with an experimental *E. coli* infection, coarsely (but not finely) ground wheat bran reduced the incidence of diarrhea in piglets (Molist et al., 2012).

In feed mills, different procedures for milling, processing, and pelleting are used and will continue to be used. The current awareness of the importance of feed structure is high, and further study of optimal diet form for performance and health is necessary.

7 Conclusion

This chapter has addressed several aspects of feeding that may affect pig health and well-being. It is clear that the diet has a large impact on an animal's functioning. The optimal diet on any given farm may depend on the farm health status: with a good health status, nutrient levels for maximal performance may be used. In contrast, farms with poorer sanitary conditions can use feeding management to decrease the risk of disease.

7.1 Key messages

- Piglets may be fed below their amino acid, calcium, and phosphorus requirements to decrease dietary buffering capacity and the risk for PWD.
- FI stimulation can be attained via both feeding management and dietary measures, affecting growth, body condition, and the health status of animals.
- Amino acids have physiological functions besides their role as a building block for protein synthesis.

- Fiber can be used to prevent disease, but more research is needed on the physiological and health-promoting role of different fiber fractions.
- Dietary energy and protein concentration, together with antioxidant vitamins and minerals may affect susceptibility to heat stress.
- Feed structure, comprising both chemical composition and physical characteristics, is important for gastric and intestinal health.
- Nutritionists face trade-offs between diets for optimal health and optimal performance.

8 Future trends

After the ban on growth-promoting antibiotics (Millet and Maertens, 2011), the pressure to diminish the use of prophylactic and therapeutic use of antibiotics is also growing. Hence, the search for feeding strategies to replace antibiotics and increase the resilience of animals, especially shortly after weaning, is ongoing and is expected to increase. We expect that feed and premix companies will increase their investments in research and we anticipate an increase in publications on dietary antibiotic replacers. Publication bias is a potential concern, as industry-funded papers may be an important marketing tool. With antibiotic resistance being one of the greatest challenges for humans in the decades to come, commercially sponsored research must be counterbalanced by adequate public funding for research. Besides strategies improving intestinal health, novel molecules targeting pathogens or toxins will be further developed; thanks to the use of advanced biotechnology, such as engineered peptides, nanoantibiotics, chicken- or plant-derived immunoglobulines, bacteriophages, eubiotics, or the use of new gene-editing technology (CRISPR/Cas9) (Marquardt and Li, 2018).

From the point of view of the environment as well as health, a decrease in dietary CP level and increase in the use of free amino acids can be expected, especially as the number of commercially available feed grade amino acids increases (see above). In the next decade, we expect to see additional research in the requirements and especially the health-promoting functions of specific amino acids.

As stated above, the role of fiber and feed structure in pig health needs further study. The rising environmental concern will likely lead to the increased use of byproducts – often containing higher fiber levels – in pig diets. Analytical methods are being developed to discriminate between the different fiber fractions. Therefore, we expect more research on the physiological and health-promoting effects of different fiber fractions and how to apply this in pig health management. The interactions between these fiber fractions and feed processing deserves also further attention.

The past decades have seen a great deal of work on the nutritional requirements of pigs. Most research is carried out in optimal conditions, while the on-farm health situation may be more variable. It can be anticipated that in the future, more emphasis will be placed on the search for precision feeding depending on the animal's characteristics, including its health status. In lactating sows and in gestating sows fed with electronic sow feeders, this is already partially done by adapting the FI to the energy requirements. Through the use of data and decision support tools, dietary strategies may be adapted to the animals' energy and amino acid needs to maximize both health and performance. The first step, albeit difficult, is to adapt the diet to the group of animals on the farm that shares similar conditions. The second step is to adapt to the individual animals' needs. To cope with the specific challenges, will require both accurate knowledge of the health status of the farm and knowledge about dietary characteristics. Tools will be needed to accurately determine the health status and the genetic potential of the animals on the farm and to translate this to farm-specific dietary advice in combination with adapted management practices. Scientific knowledge will be key in this development, although good feeding and feeding advice may go beyond the knowledge documented in scientific papers. It may require a transdisciplinary approach involving animal scientists, veterinarians, data scientists, farmers, and farm advisors. Developments in artificial intelligence may speed up this process.

9 References

Agyekum, A. K. and Nyachoti, C. M. 2017. Nutritional and metabolic consequences of feeding high-fiber diets to swine: A review. *Engineering* 3(5):716-725. doi: 10.1016/J.ENG.2017.03.010.

Aluko, K., Velayudhan, D. E., Khafipour, E., Fang, L. and Nyachoti, M. 2017. Effect of chicken egg anti-F4 antibodies on performance and diarrhea incidences in enterotoxigenic Escherichia coli K88(+)-challenged piglets. *Anim. Nutr.* 3(4):353-358. doi: 10.1016/j.aninu.2017.06.006.

Bach Knudsen, K. E. 2001. The nutritional significance of "dietary fibre" analysis. *Anim. Feed Sci. Technol.* 90(1-2):3-20. doi: 10.1016/S0377-8401(01)00193-6.

Bai, M. M., Liu, H. N., Xu, K., Wen, C. Y., Yu, R., Deng, J. P. and Yin, Y. L. 2019. Use of coated Nano zinc oxide as an additive to improve the zinc excretion and intestinal morphology of growing pigs1. *J. Anim. Sci.* 97(4):1772-1783. doi: 10.1093/jas/skz065.

Barba-Vidal, E., Martin-Orue, S. M. and Castillejos, L. 2018. Review: are we using probiotics correctly in post-weaning piglets? *Animal* 12(12):2489-2498. doi: 10.1017/S1751731118000873.

Bednorz, C., Oelgeschlager, K., Kinnemann, B., Hartmann, S., Neumann, K., Pieper, R., Bethe, A., Semmler, T., Tedin, K., Schierack, P., Wieler, L. H. and Guenther, S. 2013. The broader context of antibiotic resistance: Zinc feed supplementation of piglets increases the proportion of multi-resistant Escherichia coli in vivo. *Int. J. Med. Microbiol.* 303(6-7):396-403. doi: 10.1016/j.ijmm.2013.06.004.

Bergsma, R., Kanis, E., Verstegen, M. W. A., van der Peet-Schwering, C. M. C. and Knol, E. F. 2009. Lactation efficiency as a result of body composition dynamics and feed intake in sows. *Livest. Sci.* 125(2-3):208-222. doi: 10.1016/j.livsci.2009.04.011.

Bikker, P. and Blok, M. C. 2017. *Phophorus and Calcium Requirements of Growing Pigs and Sows.*

Bikker, P. and Jongbloed, A. W. 2013. *Koper-en zinknormen voor varkens.* 1570-8616, Wageningen UR Livestock Research.

Blasco, M., Fondevila, M. and Guada, J. A. 2005. Inclusion of wheat gluten as a protein source in diets for weaned pigs. *Anim. Res.* 54(4):297-306. doi: 10.1051/animres:2005026.

Bonde, M., Rousing, T., Badsberg, J. H. and Sørensen, J. T. 2004. Associations between lying-down behaviour problems and body condition, limb disorders and skin lesions of lactating sows housed in farrowing crates in commercial sow herds. *Livest. Prod. Sci.* 87(2-3):179-187. doi: 10.1016/j.livprodsci.2003.08.005.

Bonetti, A., Tugnoli, B., Piva, A. and Grilli, E. 2021. Towards zero zinc oxide: feedingstrategies to manage post-weaning diarrhea in piglets. *Animals (Basel)* 11(3):642.

Bruininx, E. M., Binnendijk, G. P., van der Peet-Schwering, C. M., Schrama, J. W., den Hartog, L. A., Everts, H. and Beynen, A. C. 2002. Effect of creep feed consumption on individual feed intake characteristics and performance of group-housed weanling pigs. *J. Anim. Sci.* 80(6):1413-1418. doi: 10.2527/2002.8061413x.

Cabrera, R. A., Usry, J. L., Arrellano, C., Nogueira, E. T., Kutschenko, M., Moeser, A. J. and Odle, J. 2013. Effects of creep feeding and supplemental glutamine or glutamine plus glutamate (Aminogut) on pre- and post-weaning growth performance and intestinal health of piglets. *J. Anim. Sci. Biotechnol.* 4(1):29. doi: 10.1186/2049-1891-4-29.

Canh, T. T., Verstegen, M. W., Aarnink, A. J. and Schrama, J. W. 1997. Influence of dietary factors on nitrogen partitioning and composition of urine and feces of fattening pigs. *J. Anim. Sci.* 75(3):700-706. doi: 10.2527/1997.753700x.

Charette, R., Bigras-Poulin, M. and Martineau, G.-P. 1996. Body condition evaluation in sows. *Livest. Prod. Sci.* 46(2):107-115. doi: 10.1016/0301-6226(96)00022-X.

Chen, H., Mao, X., He, J., Yu, B., Huang, Z., Yu, J., Zheng, P. and Chen, D. 2013. Dietary fibre affects intestinal mucosal barrier function and regulates intestinal bacteria in weaning piglets. *Br. J. Nutr.* 110(10):1837-1848. doi: 10.1017/S0007114513001293.

Correa-Oliveira, R., Fachi, J. L., Vieira, A., Sato, F. T. and Vinolo, M. A. 2016. Regulation of immune cell function by short-chain fatty acids. *Clin. Transl. Immunol.* 5(4):e73. doi: 10.1038/cti.2016.17.

Cota, D., Proulx, K., Smith, K. A., Kozma, S. C., Thomas, G., Woods, S. C. and Seeley, R. J. 2006. Hypothalamic mTOR signaling regulates food intake. *Science* 312(5775):927-930. doi: 10.1126/science.1124147.

Cottrell, J. J., Liu, F., Hung, A. T., DiGiacomo, K., Chauhan, S. S., Leury, B. J., Furness, J. B., Celi, P. and Dunshea, F. R. 2015. Nutritional strategies to alleviate heat stress in pigs. *Anim. Prod. Sci.* 55(12):1391-1402.

Cranwell, P. D. 1985. The development of acid and pepsin (EC 3.4.23.1) secretory capacity in the pig; the effects of age and weaning. 1. Studies in anaesthetized pigs. *Br. J. Nutr.* 54(1):305-320. doi: 10.1079/bjn19850113.

Cruz, A., Håkenåsen, I. M., Skugor, A., Mydland, L. T., Åkesson, C. P., Hellestveit, S. S., Sørby, R., Press, C. M. and Øverland, M. 2019. Candida utilis yeast as a protein source for weaned piglets: effects on growth performance and digestive function. *Livest. Sci.* 226:31–39. doi: 10.1016/j.livsci.2019.06.003.

CVB. 2020. *Tabellenboek Voeding Varkens 2020.*

Danielsen, V. and Vestergaard, E.-M. 2001. Dietary fibre for pregnant sows: Effect on performance and behaviour. *Anim. Feed Sci.Technol.* 90(1–2):71–80. doi: 10.1016/S0377-8401(01)00197-3.

de Leeuw, J. A., Bolhuis, J. E., Bosch, G. and Gerrits, W. J. 2008. Effects of dietary fibre on behaviour and satiety in pigs. *Proc. Nutr. Soc.* 67(4):334–342. doi: 10.1017/S002966510800863X.

de Souza, T. C. R., Barreyro, A. A., Rubio, S. R., González, Y. M., García, K. E., Soto, J. G. G. and Mariscal-Landín, G. 2019. Growth performance, diarrhoea incidence, and nutrient digestibility in weaned piglets fed an antibiotic-free diet with dehydrated porcine plasma or potato protein concentrate. *Ann. Anim. Sci.* 19(1):159–172.

Debski, B. 2016. Supplementation of pigs diet with zinc and copper as alternative to conventional antimicrobials. *Pol. J. Vet. Sci.* 19(4):917–924. doi: 10.1515/pjvs-2016-0113.

Decaluwe, R., Maes, D., Cools, A., Wuyts, B., De Smet, S., Marescau, B., De Deyn, P. P. and Janssens, G. P. 2014. Effect of peripartal feeding strategy on colostrum yield and composition in sows. *J. Anim. Sci.* 92(8):3557–3567. doi: 10.2527/jas.2014-7612.

Decaluwe, R., Maes, D., Declerck, I., Cools, A., Wuyts, B., De Smet, S. and Janssens, G. P. 2013. Changes in back fat thickness during late gestation predict colostrum yield in sows. *Animal* 7(12):1999–2007. doi: 10.1017/S1751731113001791.

Dourmad, J. Y., Etienne, M., Prunier, A. and Noblet, J. 1994. The effect of energy and protein intake of sows on their longevity: A review. *Livest. Prod. Sci.* 40(2):87–97. doi: 10.1016/0301-6226(94)90039-6.

Eissen, J. J., Kanis, E. and Kemp, B. 2000. Sow factors affecting voluntary feed intake during lactation. *Livest. Prod. Sci.* 64(2–3):147–165. doi: 10.1016/S0301-6226(99)00153-0.

Engle, M. J. 1994. The role of soybean meal hypersensitivity in postweaning lag and diarrhea in piglets. *J. Swine Health Prod.* 2:7–10.

European Commission. 2018. Commission implementing regulation (EU) 2018/1039 of 23 July 2018 concerning the authorisation of copper(II) diacetate monohydrate, copper(II) carbonate dihydroxy monohydrate, copper(II) chloride dihydrate, copper(II) oxide, copper(II) sulphate pentahydrate, copper(II) chelate of amino acids hydrate, copper(II) chelate of protein hydrolysates, copper(II) chelate of glycine hydrate (solid) and copper(II) chelate of glycine hydrate (liquid) as feed additives for all animal species and amending Regulations (EC) no 1334/2003, (EC). *Off. J. Eur. Union.* European Union No 1230/2014 and (EU) 2016/2261479/2006 and (EU) No 349/2010 and Implementing Regulations (EU) No 269/2012:L186/183-L186/124.

Fearnside, P. M. 2002. Soybean cultivation as a threat to the environment in Brazil. *Environ. Conserv.* 28(1):23–38. doi: 10.1017/S0376892901000030.

Feng, J., Liu, X., Xu, Z. R., Lu, Y. P. and Liu, Y. Y. 2007. Effect of fermented soybean meal on intestinal morphology and digestive enzyme activities in weaned piglets. *Dig. Dis. Sci.* 52(8):1845–1850. doi: 10.1007/s10620-006-9705-0.

Friesen, K. G., Goodband, R. D., Nelssen, J. L., Blecha, F., Reddy, D. N., Reddy, P. G. and Kats, L. J. 1993. The effect of pre- and postweaning exposure to soybean meal on growth performance and on the immune response in the early-weaned pig. *J. Anim. Sci.* 71(8):2089-2098. doi: 10.2527/1993.7182089x.

German Society for Nutrition Physiology. 2006. *Empfehlungen zur Energie-und Nährstoffversorgung von Schweinen.*

Gloaguen, M., Le Floc, N., Brossard, L., Barea, R., Primot, Y., Corrent, E. and van Milgen, J. 2011. Response of piglets to the valine content in diet in combination with the supply of other branched-chain amino acids. *Animal* 5(11):1734-1742. doi: 10.1017/S1751731111000760.

Gloaguen, M., Le Floc, N., Corrent, E., Primot, Y. and van Milgen, J. 2014. The use of free amino acids allows formulating very low crude protein diets for piglets. *J. Anim. Sci.* 92(2):637-644. doi: 10.2527/jas.2013-6514.

Goodband, B., Tokach, M., Dritz, S., Derouchey, J. and Woodworth, J. 2014. Practical starter pig amino acid requirements in relation to immunity, gut health and growth performance. *J. Anim. Sci. Biotechnol.* 5(1):12. doi: 10.1186/2049-1891-5-12.

Goodband, R. D., DeRouchey, J. M., Dritz, S. S., Woodworth, J. C., Tokach, M. D., and Cemin, H. S. 2019. Branched-chain amino acid interactions in growing pig diets1. *Translational. Animal. Science* 3(4):1246--1253. doi: 10.1093/tas/txz087.

Goodband, R. D., Tokach, M. D., Goncalves, M. A. D., Woodworth, J. C., Dritz, S. S. and DeRouchey, J. M. 2013. Nutritional enhancement during pregnancy and its effects on reproduction in swine. *Anim. Front.* 3(4):68-75. doi: 10.2527/af.2013-0036.

Guevarra, R. B., Lee, J. H., Lee, S. H., Seok, M. J., Kim, D. W., Kang, B. N., Johnson, T. J., Isaacson, R. E. and Kim, H. B. 2019. Piglet gut microbial shifts early in life: Causes and effects. *J. Anim. Sci. Biotechnol.* 10:1. doi: 10.1186/s40104-018-0308-3.

Halas, D., Hansen, C. F., Hampson, D. J., Mullan, B. P., Wilson, R. H. and Pluske, J. R. 2009. Effect of dietary supplementation with inulin and/or benzoic acid on the incidence and severity of post-weaning diarrhoea in weaner pigs after experimental challenge with enterotoxigenic Escherichia coli. *Arch. Anim. Nutr.* 63(4):267-280. doi: 10.1080/17450390903020414.

Hamard, A., Mazurais, D., Boudry, G., Le Huerou-Luron, I., Seve, B. and Le Floc, N. 2010. A moderate threonine deficiency affects gene expression profile, paracellular permeability and glucose absorption capacity in the ileum of piglets. *J. Nutr. Biochem.* 21(10):914-921. doi: 10.1016/j.jnutbio.2009.07.004.

Hansen, C. F., Riis, A. L., Bresson, S., Højbjerg, O. and Jensen, B. B. 2007. Feeding organic acids enhances the barrier function against pathogenic bacteria of the piglet stomach. *Livest. Sci.* 108(1-3):206-209. doi: 10.1016/j.livsci.2007.01.059.

He, B., Bai, Y., Jiang, L., Wang, W., Li, T., Liu, P., Tao, S., Zhao, J., Han, D. and Wang, J. 2018. Effects of oat bran on nutrient digestibility, intestinal microbiota, and inflammatory responses in the hindgut of growing pigs. *Int. J. Mol. Sci.* 19(8):2407.

He, Y., Deen, J., Shurson, G. C., Wang, L., Chen, C., Keisler, D. H. and Li, Y. Z. 2016. Identifying factors contributing to slow growth in pigs. *J. Anim. Sci.* 94(5):2103-2116. doi: 10.2527/jas.2015-0005.

Hedemann, M. S., Eskildsen, M., Laerke, H. N., Pedersen, C., Lindberg, J. E., Laurinen, P. and Knudsen, K. E. 2006. Intestinal morphology and enzymatic activity in newly weaned pigs fed contrasting fiber concentrations and fiber properties. *J. Anim. Sci.* 84(6):1375-1386. doi: 10.2527/2006.8461375x.

Hedemann, M. S., Mikkelsen, L. L., Naughton, P. J. and Jensen, B. B. 2005. Effect of feed particle size and feed processing on morphological characteristics in the small and large intestine of pigs and on adhesion of *Salmonella enterica* serovar

Typhimurium DT12 in the ileum in vitro. *J. Anim. Sci.* 83(7):1554-1562. doi: 10.2527/2005.8371554x.

Henry, Y. 1985. Dietary factors involved in feed intake regulation in growing pigs: A review. *Livest. Prod. Sci.* 12(4):339-354. doi: 10.1016/0301-6226(85)90133-2.

Heo, J. M., Kim, J. C., Hansen, C. F., Mullan, B. P., Hampson, D. J. and Pluske, J. R. 2009. Feeding a diet with decreased protein content reduces indices of protein fermentation and the incidence of postweaning diarrhea in weaned pigs challenged with an enterotoxigenic strain of Escherichia coli. *J. Anim. Sci.* 87(9):2833-2843. doi: 10.2527/jas.2008-1274.

Holen, J. P., Rambo, Z., Hilbrands, A. M. and Johnston, L. J. 2018. Effects of dietary zinc source and concentration on performance of growing-finishing pigs reared with reduced floor space. *Prof. Anim. Sci.* 34(2):133-143.

Hollis, G. R., Carter, S. D., Cline, T. R., Crenshaw, T. D., Cromwell, G. L., Hill, G. M., Kim, S. W., Lewis, A. J., Mahan, D. C., Miller, P. S., Stein, H. H., Veum, T. L. and NCRCoS. 2005. Effects of replacing pharmacological levels of dietary zinc oxide with lower dietary levels of various organic zinc sources for weanling pigs. *J. Anim. Sci.* 83(9):2123-2129. doi: 10.2527/2005.8392123x.

Hopwood, D. E., Pethick, D. W., Pluske, J. R. and Hampson, D. J. 2004. Addition of pearl barley to a rice-based diet for newly weaned piglets increases the viscosity of the intestinal contents, reduces starch digestibility and exacerbates post-weaning colibacillosis. *Br. J. Nutr.* 92(3):419-427. doi: 10.1079/bjn20041206.

Htoo, J. K., Araiza, B. A., Sauer, W. C., Rademacher, M., Zhang, Y., Cervantes, M. and Zijlstra, R. T. 2007. Effect of dietary protein content on ileal amino acid digestibility, growth performance, and formation of microbial metabolites in ileal and cecal digesta of early-weaned pigs. *J. Anim. Sci.* 85(12):3303-3312. doi: 10.2527/jas.2007-0105.

Hu, C., Song, J., Li, Y., Luan, Z. and Zhu, K. 2013. Diosmectite-zinc oxide composite improves intestinal barrier function, modulates expression of pro-inflammatory cytokines and tight junction protein in early weaned pigs. *Br. J. Nutr.* 110(4):681-688. doi: 10.1017/S0007114512005508.

Isabel, B., Rey, A. I. and López Bote, C. 2012. *Optimum Vitamin Nutrition in Pigs, Optimum Vitamin Nutrition in the Production of Quality Animal Foods. 5M Publishing.* Sheffield, pp. 243-308.

Jansman, A., van Diepen, J., Rovers, M. and Corrent, E. 2016. *Lowering the Dietary Protein Content in Piglets: How Far Can We Go?, Energy and Protein Metabolism and Nutrition,* pp. 161-162.

Jayaraman, B., Htoo, J. and Nyachoti, C. M. 2015. Effects of dietary threonine:lysine ratioes and sanitary conditions on performance, plasma urea nitrogen, plasma-free threonine and lysine of weaned pigs. *Anim. Nutr.* 1(4):283-288. doi: 10.1016/j.aninu.2015.09.003.

Jha, R. and Berrocoso, J. F. D. 2016. Dietary fiber and protein fermentation in the intestine of swine and their interactive effects on gut health and on the environment: A review. *Anim. Feed Sci. Technol.* 212:18-26. doi: 10.1016/j.anifeedsci.2015.12.002.

Jha, R., Fouhse, J. M., Tiwari, U. P., Li, L. and Willing, B. P. 2019. Dietary fiber and intestinal health of monogastric animals. *Front. Vet. Sci.* 6:48. doi: 10.3389/fvets.2019.00048.

Ji, F. J., Wang, L. X., Yang, H. S., Hu, A. and Yin, Y. L. 2019. Review: the roles and functions of glutamine on intestinal health and performance of weaning pigs. *Animal* 13(11):2727-2735. doi: 10.1017/S1751731119001800.

Johnson, J. S. and Lay, D. C. 2017. Evaluating the behavior, growth performance, immune parameters, and intestinal morphology of weaned piglets after simulated transport and heat stress when antibiotics are eliminated from the diet or replaced with L-glutamine. *J. Anim. Sci.* 95(1):91–102.

Jondreville, C., Revy, P. S. and Dourmad, J. Y. 2003. Dietary means to better control the environmental impact of copper and zinc by pigs from weaning to slaughter. *Livest. Prod. Sci.* 84(2):147–156. doi: 10.1016/j.livprodsci.2003.09.011.

Jongbloed, A., van Diepen, J. T. M., Kemme, P. and Veevoederbureau, C. 2003. *Fosfornormen voor varkens: Herziening 2003. Centraal Veevoederbureau.*

Kiarie, E., Romero, L. F. and Nyachoti, C. M. 2013. The role of added feed enzymes in promoting gut health in swine and poultry. *Nutr. Res. Rev.* 26(1):71–88. doi: 10.1017/S0954422413000048.

Kiers, J. L., Meijer, J. C., Nout, M. J., Rombouts, F. M., Nabuurs, M. J. and van der Meulen, J. 2003. Effect of fermented soya beans on diarrhoea and feed efficiency in weaned piglets. *J. Appl. Microbiol.* 95(3):545–552. doi: 10.1046/j.1365-2672.2003.02011.x.

KilBride, A. L., Gillman, C. E. and Green, L. E. 2009. A cross sectional study of the prevalence, risk factors and population attributable fractions for limb and body lesions in lactating sows on commercial farms in England. *BMC Vet. Res.* 5:30. doi: 10.1186/1746-6148-5-30.

Kloubert, V., Blaabjerg, K., Dalgaard, T. S., Poulsen, H. D., Rink, L. and Wessels, I. 2018. Influence of zinc supplementation on immune parameters in weaned pigs. *J. Trace Elem. Med. Biol.* 49:231–240. doi: 10.1016/j.jtemb.2018.01.006.

Konowalchuk, J. D., Rieger, A. M., Kiemele, M. D., Ayres, D. C. and Barreda, D. R. 2013. Modulation of weanling pig cellular immunity in response to diet supplementation with 25-hydroxyvitamin D(3). *Vet. Immunol. Immunopathol.* 155(1–2):57–66. doi: 10.1016/j.vetimm.2013.06.002.

Lacey, J. M. and Wilmore, D. W. 1990. Is glutamine a conditionally essential amino acid? *Nutr. Rev.* 48(8):297–309. doi: 10.1111/j.1753-4887.1990.tb02967.x.

Lallès, J.-P. and Montoya, C. A. 2021. Dietary alternatives to in-feed antibiotics, gut barrier function and inflammation in piglets post-weaning: Where are we now? *Anim. Feed Sci. Technol.* 274:274. doi: 10.1016/j.anifeedsci.2021.114836.

Lauridsen, C. 2019. From oxidative stress to inflammation: Redox balance and immune system. *Poult. Sci.* 98(10):4240–4246. doi: 10.3382/ps/pey407.

Lauridsen, C., Matte, J. J., Lessard, M., Celi, P. and Litta, G. 2021. Role of vitamins for gastro-intestinal functionality and health of pigs. *Anim. Feed Sci. Technol.* 273:273. doi: 10.1016/j.anifeedsci.2021.114823.

Lawlor, P. G., Lynch, P. B., Caffrey, P. J., O'Reilly, J. J. and O'Connell, M. K. 2005. Measurements of the acid-binding capacity of ingredients used in pig diets. *Ir. Vet. J.* 58(8):447–452. doi: 10.1186/2046-0481-58-8-447.

Le Floc'h, N., Wessels, A., Corrent, E., Wu, G. and Bosi, P. 2018. The relevance of functional amino acids to support the health of growing pigs. *Anim. Feed Sci. Technol.* 245:104–116. doi: 10.1016/j.anifeedsci.2018.09.007.

Leal, D. F., Muro, B. B. D., Nichi, M., Almond, G. W., Viana, C. H. C., Vioti, G., Carnevale, R. F. and Garbossa, C. A. P. 2019. Effects of post-insemination energy content of feed on embryonic survival in pigs: A systematic review. *Anim. Reprod. Sci.* 205:70–77. doi: 10.1016/j.anireprosci.2019.04.005.

Lessard, M., Dupuis, M., Gagnon, N., Nadeau, E., Matte, J. J., Goulet, J. and Fairbrother, J. M. 2009. Administration of Pediococcus acidilactici or Saccharomyces cerevisiae boulardii modulates development of porcine mucosal immunity and reduces intestinal bacterial translocation after Escherichia coli challenge. *J. Anim. Sci.* 87(3):922-934. doi: 10.2527/jas.2008-0919.

Li, P., Yin, Y. L., Li, D., Kim, S. W. and Wu, G. 2007. Amino acids and immune function. *Br. J. Nutr.* 98(2):237-252. doi: 10.1017/S000711450769936X.

Liedtke, J. and Vahjen, W. 2012. In vitro antibacterial activity of zinc oxide on a broad range of reference strains of intestinal origin. *Vet. Microbiol.* 160(1-2):251-255. doi: 10.1016/j.vetmic.2012.05.013.

Lillehoj, H., Liu, Y., Calsamiglia, S., Fernandez-Miyakawa, M. E., Chi, F., Cravens, R. L., Oh, S. and Gay, C. G. 2018. Phytochemicals as antibiotic alternatives to promote growth and enhance host health. *Vet. Res.* 49(1):76. doi: 10.1186/s13567-018-0562-6.

Lisgara, M, Skampardonis, V. and Leontides, L. 2016. Effect of diet supplementation with chelated zinc, copper and manganese on hoof lesions of loose housed sows. *Porcine Health Manag.* 2:6. doi: 10.1186/s40813-016-0025-2.

Liu, F., Cottrell, J. J., Furness, J. B., Rivera, L. R., Kelly, F. W., Wijesiriwardana, U., Pustovit, R. V., Fothergill, L. J., Bravo, D. M., Celi, P., Leury, B. J., Gabler, N. K. and Dunshea, F. R. 2016. Selenium and vitamin E together improve intestinal epithelial barrier function and alleviate oxidative stress in heat-stressed pigs. *Exp. Physiol.* 101(7):801-810. doi: 10.1113/EP085746.

Liu, Y., Huang, J., Hou, Y., Zhu, H., Zhao, S., Ding, B., Yin, Y., Yi, G., Shi, J. and Fan, W. 2008. Dietary arginine supplementation alleviates intestinal mucosal disruption induced by Escherichia coli lipopolysaccharide in weaned pigs. *Br. J. Nutr.* 100(3):552-560. doi: 10.1017/S0007114508911612.

Liu, Y., Wang, X., Hou, Y., Yin, Y., Qiu, Y., Wu, G. and Hu, C. A. 2017. Roles of amino acids in preventing and treating intestinal diseases: Recent studies with pig models. *Amino Acids* 49(8):1277-1291. doi: 10.1007/s00726-017-2450-1.

Lorenzett, M. P., Cruz, R. A. S., Cecco, B. S., Schwertz, C. I., Hammerschmitt, M. E., Schu, D. T., Driemeier, D. and Pavarini, S. P. 2019. Obstructive urolithiasis in growing-finishing pigs. *Pesq. Vet. Bras.* 39(6):382-387. doi: 10.1590/1678-5150-pvb-6229.

Lupton, J. R., Betteridge, V. A. and Pijls, L. T. J. 2009. Codex final definition of dietary fibre: Issues of implementation. *Qual. Assur. Saf. Crops Foods* 1(4):206-212. doi: 10.1111/j.1757-837X.2009.00043.x.

Maes, D. G., Vrielinck, J., Millet, S., Janssens, G. P. and Deprez, P. 2004a. Urolithiasis in finishing pigs. *Vet. J.* 168(3):317-322. doi: 10.1016/j.tvjl.2003.09.006.

Maes, D. G. D., Janssens, G. P. J., Delputte, P., Lammertyn, A. and de Kruif, A. 2004b. Back fat measurements in sows from three commercial pig herds: Relationship with reproductive efficiency and correlation with visual body condition scores. *Livest. Prod. Sci.* 91(1-2):57-67. doi: 10.1016/j.livprodsci.2004.06.015.

Mahan, D. C. and Shields, R. G., Jr. 1998. Essential and nonessential amino acid composition of pigs from birth to 145 kilograms of body weight, and comparison to other studies. *J. Anim. Sci.* 76(2):513-521. doi: 10.2527/1998.762513x.

Mallmann, A. L., Oliveira, G. S., Ulguim, R. R., Mellagi, A. P. G., Bernardi, M. L., Orlando, U. A. D., Goncalves, M. A. D., Cogo, R. J. and Bortolozzo, F. P. 2020. Impact of feed intake in early gestation on maternal growth and litter size according to body reserves at weaning of young parity sows. *J. Anim. Sci.* 98(3).

Markowiak, P. and Slizewska, K. 2018. The role of probiotics, prebiotics and Synbiotics in animal nutrition. *Gut Pathog.* 10:21. doi: 10.1186/s13099-018-0250-0.

Marquardt, R. R., Jin, L. Z., Kim, J. W., Fang, L., Frohlich, A. A. and Baidoo, S. K. 1999. Passive protective effect of egg-yolk antibodies against enterotoxigenic Escherichia coli K88+ infection in neonatal and early-weaned piglets. *FEMS Immunol. Med. Microbiol.* 23(4):283–288. doi: 10.1111/j.1574-695X.1999.tb01249.x.

Marquardt, R. R. and Li, S. 2018. Antimicrobial resistance in livestock: Advances and alternatives to antibiotics. *Anim. Front.* 8(2):30–37. doi: 10.1093/af/vfy001.

Mavromatis, J., Koptopoulos, G., Kyriakis, S. C., Papasteriadis, A. and Saoulidis, K. 1999. Effects of alpha-tocopherol and selenium on pregnant sows and their piglets' immunity and performance. *Zentralbl. Veterinarmed. A* 46(9):545–553. doi: 10.1046/j.1439-0442.1999.00244.x.

Maxwell, C. V., Reimann, E. M., Hoekstra, W. G., Kowalczyk, T., Benevenga, N. J. and Grummer, R. H. 1970. Effect of dietary particle size on lesion development and on the contents of various regions of the swine stomach. *J. Anim. Sci.* 30(6):911–922. doi: 10.2527/jas1970.306911x.

McDonald, D. E., Pethick, D. W., Pluske, J. R. and Hampson, D. J. 1999. Adverse effects of soluble non-starch polysaccharide (guar gum) on piglet growth and experimental colibacillosis immediately after weaning. *Res. Vet. Sci.* 67(3):245–250. doi: 10.1053/rvsc.1999.0315.

Metzler-Zebeli, B. U., Canibe, N., Montagne, L., Freire, J., Bosi, P., Prates, J. A. M., Tanghe, S. and Trevisi, P. 2019. Resistant starch reduces large intestinal pH and promotes fecal lactobacilli and bifidobacteria in pigs. *Animal* 13(1):64–73. doi: 10.1017/S1751731118001003.

Mikkelsen, L. L., Naughton, P. J., Hedemann, M. S. and Jensen, B. B. 2004. Effects of physical properties of feed on microbial ecology and survival of Salmonella enterica serovar Typhimurium in the pig gastrointestinal tract. *Appl. Environ. Microbiol.* 70(6):3485–3492. doi: 10.1128/AEM.70.6.3485-3492.2004.

Milani, N. C., Sbardella, M., Ikeda, N. Y., Arno, A., Mascarenhas, B. C. and Miyada, V. S. 2017. Dietary zinc oxide nanoparticles as growth promoter for weanling pigs. *Anim. Feed Sci. Technol.* 227:13–23. doi: 10.1016/j.anifeedsci.2017.03.001.

Millet, S., Aluwe, M., Ampe, B. and De Campeneere, S. 2015. Interaction between amino acids on the performances of individually housed piglets. *J. Anim. Physiol. Anim. Nutr. (Berl)* 99(2):230–236. doi: 10.1111/jpn.12227.

Millet, S., Aluwe, M., De Boever, J., De Witte, B., Douidah, L., Van den Broeke, A., Leen, F., De Cuyper, C., Ampe, B. and De Campeneere, S. 2018. The effect of crude protein reduction on performance and nitrogen metabolism in piglets (four to nine weeks of age) fed two dietary lysine levels1. *J. Anim. Sci.* 96(9):3824–3836. doi: 10.1093/jas/sky254.

Millet, S., Kumar, S., De Boever, J., Ducatelle, R. and De Brabander, D. 2012a. Effect of feed processing on growth performance and gastric mucosa integrity in pigs from weaning until slaughter. *Anim. Feed Sci. Technol.* 175(3-4):175–181. doi: 10.1016/j.anifeedsci.2012.05.010.

Millet, S., Kumar, S., De Boever, J., Meyns, T., Aluwe, M., De Brabander, D. and Ducatelle, R. 2012b. Effect of particle size distribution and dietary crude fibre content on growth performance and gastric mucosa integrity of growing-finishing pigs. *Vet. J.* 192(3):316–321. doi: 10.1016/j.tvjl.2011.06.037.

Millet, S. and Maertens, L. 2011. The European ban on antibiotic growth promoters in animal feed: From challenges to opportunities. *Vet. J.* 187(2):143–144. doi: 10.1016/j.tvjl.2010.05.001.

Millet, S., Chalvon-Demersay, T., Schroyen, M., De Cuyper, C., Aluwé, M., Ampe, B., Lambert, W., Bannai, M., Yasuhiro, O. and Simongiovanni, A. 2020a. La performance des porcelets au sevrage répond à une supplémentation en glutamine, en particulier avec un régime faible en protéines. *J. Recherche porcine* 52:167–168.

Millet, S., Minussi, I., Lambert, W., Aluwé, M., Ampe, B., De Sutter, J. and De Campeneere, S. 2020b. Standardized ileal digestible lysine and valine-to-lysine requirements for optimal performance of 4 to 9-week-old Piétrain cross piglets. *Livest. Sci.* 241:241. doi: 10.1016/j.livsci.2020.104263.

Misiura, M. M., Filipe, J. A. N., Walk, C. L. and Kyriazakis, I. 2020. How do pigs deal with dietary phosphorus deficiency? *Br. J. Nutr.* 124(3):256–272. doi: 10.1017/S0007114520000975.

Molist, F., Manzanilla, E. G., Perez, J. F. and Nyachoti, C. M. 2012. Coarse, but not finely ground, dietary fibre increases intestinal Firmicutes:Bacteroidetes ratio and reduces diarrhoea induced by experimental infection in piglets. *Br. J. Nutr.* 108(1):9–15. doi: 10.1017/S0007114511005216.

Montagne, L., Cavaney, F. S., Hampson, D. J., Lalles, J. P. and Pluske, J. R. 2004. Effect of diet composition on postweaning colibacillosis in piglets. *J. Anim. Sci.* 82(8):2364–2374. doi: 10.2527/2004.8282364x.

Navarro, D. M. D. L., Abelilla, J. J. and Stein, H. H. 2019. Structures and characteristics of carbohydrates in diets fed to pigs: A review. *J. Anim. Sci. Biotechnol.* 10:39. doi: 10.1186/s40104-019-0345-6.

Noblet, J., Fortune, H., Shi, X. S. and Dubois, S. 1994. Prediction of net energy value of feeds for growing pigs. *J. Anim. Sci.* 72(2):344–354. doi: 10.2527/1994.722344x.

NRC. 2012. *National Research Council. Nutrient Requirements of Swine* (11th Rev. Edn.). Washington, DC: National Academy Press.

Nyachoti, C. M., Omogbenigun, F. O., Rademacher, M. and Blank, G. 2006. Performance responses and indicators of gastrointestinal health in early-weaned pigs fed low-protein amino acid-supplemented diets. *J. Anim. Sci.* 84(1):125–134. doi: 10.2527/2006.841125x.

Oliviero, C., Heinonen, M., Valros, A. and Peltoniemi, O. 2010. Environmental and sow-related factors affecting the duration of farrowing. *Anim. Reprod. Sci.* 119(1–2):85–91. doi: 10.1016/j.anireprosci.2009.12.009.

Omonijo, F. A., Ni, L., Gong, J., Wang, Q., Lahaye, L. and Yang, C. 2018. Essential oils as alternatives to antibiotics in swine production. *Anim. Nutr.* 4(2):126–136. doi: 10.1016/j.aninu.2017.09.001.

Oostindjer, M., Bolhuis, J. E., van den Brand, H., Roura, E. and Kemp, B. 2010. Prenatal flavor exposure affects growth, health and behavior of newly weaned piglets. *Physiol. Behav.* 99(5):579–586. doi: 10.1016/j.physbeh.2010.01.031.

Pearce, S. C., Sanz Fernandez, M. V., Torrison, J., Wilson, M. E., Baumgard, L. H. and Gabler, N. K. 2015. Dietary organic zinc attenuates heat stress-induced changes in pig intestinal integrity and metabolism. *J. Anim. Sci.* 93(10):4702–4713. doi: 10.2527/jas.2015-9018.

Pedersen, C. and Lindberg, J. E. 2004. Comparison of low-glycoalkaloid potato protein and fish meal as protein sources for weaner piglets. *Acta Agric. Scand. A* 54(2):75–80.

Peeters, L., Mostin, L., Wattiau, P., Boyen, F., Dewulf, J. and Maes, D. 2019. Efficacy of Clostridium butyricum as probiotic feed additive against experimental Salmonella Typhimurium infection in pigs. *Livest. Sci.* 221:82–85. doi: 10.1016/j.livsci.2018.12.019.

Pei, X., Xiao, Z., Liu, L., Wang, G., Tao, W., Wang, M., Zou, J. and Leng, D. 2019. Effects of dietary zinc oxide nanoparticles supplementation on growth performance, zinc status, intestinal morphology, microflora population, and immune response in weaned pigs. *J. Sci. Food Agric.* 99(3):1366–1374. doi: 10.1002/jsfa.9312.

Peltoniemi, O. A. T., Oliviero, C., Hälli, O. and Heinonen, M. 2007. Feeding affects reproductive performance and reproductive endocrinology in the gilt and sow. *Acta Vet. Scand.* 49(Suppl 1).

Perez-Bosque, A., Polo, J. and Torrallardona, D. 2016. Spray dried plasma as an alternative to antibiotics in piglet feeds, mode of action and biosafety. *Porcine Health Manag.* 2:16. doi: 10.1186/s40813-016-0034-1.

Pinelli-Saavedra, A., Calderon de la Barca, A. M., Hernandez, J., Valenzuela, R. and Scaife, J. R. 2008. Effect of supplementing sows' feed with alpha-tocopherol acetate and vitamin C on transfer of alpha-tocopherol to piglet tissues, colostrum, and milk: Aspects of immune status of piglets. *Res. Vet. Sci.* 85(1):92–100. doi: 10.1016/j.rvsc.2007.08.007.

Pluske, J. R., Siba, P. M., Pethick, D. W., Durmic, Z., Mullan, B. P. and Hampson, D. J. 1996. The incidence of swine dysentery in pigs can be reduced by feeding diets that limit the amount of fermentable substrate entering the large intestine. *J. Nutr.* 126(11):2920–2933. doi: 10.1093/jn/126.11.2920.

Pluym, L. M., Van Nuffel, A., Van Weyenberg, S. and Maes, D. 2013. Prevalence of lameness and claw lesions during different stages in the reproductive cycle of sows and the impact on reproduction results. *Animal* 7(7):1174–1181. doi: 10.1017/S1751731113000232.

Renaudeau, D., Gourdine, J., Silva, B. and Noblet, J. 2008. *Nutritional Routes to Attenuate Heat Stress in Pigs, Livestock and Global Climate Change*. Cambridge: Cambridge University Press, pp. 134–138.

Resnik, D. B. 2000. Financial interests and research bias. *Perspect. Sci.* 8(3):255–285. doi: 10.1162/106361400750340497.

Richards, J. D., Zhao, J., Harrell, R. J., Atwell, C. A. and Dibner, J. J. 2010. Trace mineral nutrition in poultry and swine. *Asian-Australas. J. Anim. Sci.* 23(11):1527–1534.

Rioja-Lang, F. C., Seddon, Y. M. and Brown, J. A. 2018. Shoulder lesions in sows: A review of their causes, prevention, and treatment. *J. Swine Health Prod.* 26(2):101–107.

Rist, V. T., Weiss, E., Eklund, M. and Mosenthin, R. 2013. Impact of dietary protein on microbiota composition and activity in the gastrointestinal tract of piglets in relation to gut health: A review. *Animal* 7(7):1067–1078. doi: 10.1017/S1751731113000062.

Rodrigues, L. A., Wellington, M. O., Gonzalez-Vega, J. C., Htoo, J. K., Van Kessel, A. G. and Columbus, D. A. 2021. Functional amino acid supplementation, regardless of dietary protein content, improves growth performance and immune status of weaned pigs challenged with Salmonella Typhimurium. *J. Anim. Sci.* 99(2):skaa365.

Schlegel, P., Sauvant, D. and Jondreville, C. 2013. Bioavailability of zinc sources and their interaction with phytates in broilers and piglets. *Animal* 7(1):47–59. doi: 10.1017/S1751731112001000.

Siba, P. M., Pethick, D. W. and Hampson, D. J. 1996. Pigs experimentally infected with *Serpulina hyodysenteriae* can be protected from developing swine dysentery by feeding them a highly digestible diet. *Epidemiol. Infect.* 116(2):207–216. doi: 10.1017/s0950268800052456.

Slavin, J. 2013. Fiber and prebiotics: Mechanisms and health benefits. *Nutrients* 5(4):1417–1435. doi: 10.3390/nu5041417.

Solà-Oriol, D. and Gasa, J. 2017. Feeding strategies in pig production: sows and their piglets. *Anim. Feed Sci. Technol.* 233:34–52. doi: 10.1016/j.anifeedsci.2016.07.018.

Song, Z. H., Xiao, K., Ke, Y. L., Jiao, le F. and Hu, C. H. 2015. Zinc oxide influences mitogen-activated protein kinase and TGF-beta1 signaling pathways, and enhances intestinal barrier integrity in weaned pigs. *Innate Immun.* 21(4):341–348. doi: 10.1177/1753425914536450.

Sun, Y. and Kim, S. W. 2017. Intestinal challenge with enterotoxigenic Escherichia coli in pigs, and nutritional intervention to prevent postweaning diarrhea. *Anim. Nutr.* 3(4):322–330. doi: 10.1016/j.aninu.2017.10.001.

Sun, Y., Zhou, L., Fang, L., Su, Y. and Zhu, W. 2015. Responses in colonic microbial community and gene expression of pigs to a long-term high resistant starch diet. *Front. Microbiol.* 6:877. doi: 10.3389/fmicb.2015.00877.

Święch, E., Tuśnio, A., Taciak, M., Boryczka, M. and Buraczewska, L. 2012. The effects of pectin and rye on amino acid ileal digestibility, threonine metabolism, nitrogen retention, and morphology of the small intestine in young pigs. *J. Anim. Feed Sci.* 21(1):89–106. doi: 10.22358/jafs/66055/2012.

Trevisi, P., Corrent, E., Mazzoni, M., Messori, S., Priori, D., Gherpelli, Y., Simongiovanni, A. and Bosi, P. 2015. Effect of added dietary threonine on growth performance, health, immunity and gastrointestinal function of weaning pigs with differing genetic susceptibility to Escherichia coli infection and challenged with E. coli K88ac. *J. Anim. Physiol. Anim. Nutr. (Berl)* 99(3):511–520. doi: 10.1111/jpn.12216.

Trevisi, P., Latorre, R., Priori, D., Luise, D., Archetti, I., Mazzoni, M., D'Inca, R. and Bosi, P. 2017. Effect of feed supplementation with live yeast on the intestinal transcriptome profile of weaning pigs orally challenged with Escherichia coli F4. *Animal* 11(1):33–44. doi: 10.1017/S1751731116001178.

Tucker, H. F. and Salmon, W. D. 1955. Parakeratosis or zinc deficiency disease in the pig. *Proc. Soc. Exp. Biol. Med.* 88(4):613–616.

Tugnoli, B., Giovagnoni, G., Piva, A. and Grilli, E. 2020. From acidifiers to intestinal health enhancers: How organic acids can improve growth efficiency of pigs. *Animals (Basel)* 10(1), 134. doi: 10.3390/ani10010134.

van der Meer, Y., Lammers, A., Jansman, A. J., Rijnen, M. M., Hendriks, W. H. and Gerrits, W. J. 2016. Performance of pigs kept under different sanitary conditions affected by protein intake and amino acid supplementation. *J. Anim. Sci.* 94(11):4704–4719. doi: 10.2527/jas.2016-0787.

van Heugten, E., Spears, J. W., Kegley, E. B., Ward, J. D. and Qureshi, M. A. 2003. Effects of organic forms of zinc on growth performance, tissue zinc distribution, and immune response of weanling pigs. *J. Anim. Sci.* 81(8):2063–2071. doi: 10.2527/2003.8182063x.

van Riet, M. M., Janssens, G. P., Cornillie, P., Van Den Broeck, W., Nalon, E., Ampe, B., Tuyttens, F. A., Maes, D., Du Laing, G. and Millet, S. 2016a. Marginal dietary zinc concentration affects claw conformation measurements but not histological claw characteristics in weaned pigs. *Vet. J.* 209:98–107. doi: 10.1016/j.tvjl.2016.01.007.

van Riet, M. M., Millet, S., Liesegang, A., Nalon, E., Ampe, B., Tuyttens, F. A., Maes, D. and Janssens, G. P. 2016b. Impact of parity on bone metabolism throughout the reproductive cycle in sows. *Animal* 10(10):1714–1721. doi: 10.1017/S1751731116000471.

van Riet, M. M. J., Millet, S., Aluwé, M. and Janssens, G. P. J. 2013. Impact of nutrition on lameness and claw health in sows. *Livest. Sci.* 156(1–3):24–35. doi: 10.1016/j.livsci.2013.06.005.

van Zanten, H. H., Bikker, P., Mollenhorst, H., Meerburg, B. G. and de Boer, I. J. 2015. Environmental impact of replacing soybean meal with rapeseed meal in diets of finishing pigs. *Animal* 9(11):1866–1874. doi: 10.1017/S1751731115001469.

Vente-Spreeuwenberg, M. A. M., Verdonk, J. M. A. J., Bakker, G. C. M., Beynen, A. C. and Verstegen, M. W. A. 2004. Effect of dietary protein source on feed intake and small intestinal morphology in newly weaned piglets. *Livest. Prod. Sci.* 86(1–3):169–177. doi: 10.1016/S0301-6226(03)00166-0.

Vrielinck, J., Sarrazin, S., Schoos, A., Janssens, G. P. J. and Maes, D. 2019. Prevalence and chemical composition of uroliths in fattening pigs in Belgium. *J. Anim. Physiol. Anim. Nutr. (Berl)* 103(6):1828–1836. doi: 10.1111/jpn.13169.

Vukmirović, Đ., Čolović, R., Rakita, S., Brlek, T., Đuragić, O. and Solà-Oriol, D. 2017. Importance of feed structure (particle size) and feed form (mash vs. pellets) in pig nutrition - A review. *Anim. Feed Sci.Technol.* 233:133–144. doi: 10.1016/j.anifeedsci.2017.06.016.

Wang, C., Zhang, L., Ying, Z., He, J., Zhou, L., Zhang, L., Zhong, X. and Wang, T. 2018. Effects of dietary zinc oxide nanoparticles on growth, diarrhea, mineral deposition, intestinal morphology, and barrier of weaned piglets. *Biol. Trace Elem. Res.* 185(2):364–374. doi: 10.1007/s12011-018-1266-5.

Wang, L., Xu, X., Su, G., Shi, B. and Shan, A. 2017. High concentration of vitamin E supplementation in sow diet during the last week of gestation and lactation affects the immunological variables and antioxidative parameters in piglets. *J. Dairy Res.* 84(1):8–13. doi: 10.1017/S0022029916000650.

Wang, W., Wang, Y., Hao, X., Duan, Y., Meng, Z., An, X. and Qi, J. 2020. Dietary fermented soybean meal replacement alleviates diarrhea in weaned piglets challenged with enterotoxigenic Escherichia coli K88 by modulating inflammatory cytokine levels and cecal microbiota composition. *BMC Vet. Res.* 16(1):245. doi: 10.1186/s12917-020-02466-5.

Wang, X., Feng, Y., Shu, G., Jiang, Q., Yang, J. and Zhang, Z. 2011. Effect of dietary supplementation with hydrolyzed wheat gluten on growth performance, cell immunity and serum biochemical indices of weaned piglets (Sus scrofa). *Agric. Sci. China* 10(6):938–945.

Wang, Z., Li, J., Li, J., Li, Y., Wang, L., Wang, Q., Fang, L., Ding, X., Huang, P., Yin, J., Yin, Y. and Yang, H. 2019. Protective effect of chicken egg yolk immunoglobulins (IgY) against enterotoxigenic Escherichia coli K88 adhesion in weaned piglets. *BMC Vet. Res.* 15(1):234. doi: 10.1186/s12917-019-1958-x.

Wegner, B., Tenhündfeld, J., Vogels, J., Beumer, M., Kamphues, J., Hansmann, F., Rieger, H., Grosse Beilage, E. and Hennig-Pauka, I. 2020. Lameness in fattening pigs-Mycoplasma hyosynoviae, osteochondropathy and reduced dietary phosphorus level as three influencing factors: A case report. *Porcine Health Manag.* 6(1):41.

Weldon, W. C., Lewis, A. J., Louis, G. F., Kovar, J. L., Giesemann, M. A. and Miller, P. S. 1994. Postpartum hypophagia in primiparous sows: I. Effects of gestation feeding level on feed intake, feeding behavior, and plasma metabolite concentrations during lactation. *J. Anim. Sci.* 72(2):387–394. doi: 10.2527/1994.722387x.

Wellock, I. J., Fortomaris, P. D., Houdijk, J. G. and Kyriazakis, I. 2008a. Effects of dietary protein supply, weaning age and experimental enterotoxigenic Escherichia coli infection on newly weaned pigs: Performance. *Animal* 2(6):825–833. doi: 10.1017/S1751731108001559.

Wellock, I. J., Fortomaris, P. D., Houdijk, J. G., Wiseman, J. and Kyriazakis, I. 2008b. The consequences of non-starch polysaccharide solubility and inclusion level on the health and performance of weaned pigs challenged with enterotoxigenic Escherichia coli. *Br. J. Nutr.* 99(3):520–530. doi: 10.1017/S0007114507819167.

Wellock, I. J., Fortomaris, P. D., Houdijk, J. G. M. and Kyriazakis, I. 2007. The effect of dietary protein supply on the performance and risk of post-weaning enteric disorders in newly weaned pigs. *Anim. Sci.* 82(3):327–335. doi: 10.1079/ASC200643.

Wensley, M. R., Tokach, M. D., Woodworth, J. C., Goodband, R. D., Gebhardt, J. T., DeRouchey, J. M. and McKilligan, D. 2021a. Maintaining continuity of nutrient intake after weaning I: Review of pre-weaning strategies. *Transl. Anim. Sci.* 5(1):txab021. doi: 10.1093/tas/txab021.

Wensley, M. R., Tokach, M. D., Woodworth, J. C., Goodband, R. D., Gebhardt, J. T., DeRouchey, J. M. and McKilligan, D. 2021b. Maintaining continuity of nutrient intake after weaning II: Review of post-weaning strategies. *Transl. Anim. Sci.* 5(1):txab022. doi: 10.1093/tas/txab022.

Wijtten, P. J., van der Meulen, J. and Verstegen, M. W. 2011. Intestinal barrier function and absorption in pigs after weaning: A review. *Br. J. Nutr.* 105(7):967–981. doi: 10.1017/S0007114510005660.

Wolf, P., Rust, P. and Kamphues, J. 2010. How to assess particle size distribution in diets for pigs? *Livest. Sci.* 133(1–3):78–80. doi: 10.1016/j.livsci.2010.06.030.

Wu, G. 2014. Dietary requirements of synthesizable amino acids by animals: A paradigm shift in protein nutrition. *J. Anim. Sci. Biotechnol.* 5(1):34. doi: 10.1186/2049-1891-5-34.

Wu, G., Meier, S. A. and Knabe, D. A. 1996. Dietary glutamine supplementation prevents jejunal atrophy in weaned pigs. *J. Nutr.* 126(10):2578–2584. doi: 10.1093/jn/126.10.2578.

Xie, Z., Hu, L., Li, Y., Geng, S., Cheng, S., Fu, X., Zhao, S. and Han, X. 2017. Changes of gut microbiota structure and morphology in weaned piglets treated with fresh fermented soybean meal. *World J. Microbiol. Biotechnol.* 33(12):213. doi: 10.1007/s11274-017-2374-7.

Yokoyama, H., Peralta, R. C., Diaz, R., Sendo, S., Ikemori, Y. and Kodama, Y. 1992. Passive protective effect of chicken egg yolk immunoglobulins against experimental enterotoxigenic Escherichia coli infection in neonatal piglets. *Infect. Immun.* 60(3):998–1007.

Zhang, H., Chen, Y., Li, Y., Zhang, T., Ying, Z., Su, W., Zhang, L. and Wang, T. 2019. l-threonine improves intestinal mucin synthesis and immune function of intrauterine growth-retarded weanling piglets. *Nutrition* 59:182–187. doi: 10.1016/j.nut.2018.07.114.

Zheng, P., Yu, B., He, J., Yu, J., Mao, X., Luo, Y., Luo, J., Huang, Z., Tian, G., Zeng, Q., Che, L. and Chen, D. 2017. Arginine metabolism and its protective effects on intestinal health and functions in weaned piglets under oxidative stress induced by diquat. *Br. J. Nutr.* 117(11):1495–1502. doi: 10.1017/S0007114517001519.

Zurbrigg, K. 2006. Sow shoulder lesions: Risk factors and treatment effects on an Ontario farm. *J. Anim. Sci.* 84(9):2509–2514. doi: 10.2527/jas.2005-713.

Chapter 3

The use of prebiotics to optimize gut function in pigs

Barbara U. Metzler-Zebeli, University of Veterinary Medicine Vienna, Austria

1 Introduction

To survive and multiply, microorganisms rely on a supply of nutrients from their host. Using diet to manipulate the porcine gut microbiota as a way to promote gut health is a widely used strategy in pig production (Newman et al., 2018; Schokker et al., 2018). The idea of using diet to increase beneficial bacteria in the gastrointestinal tract (GIT) to control pathogens has led to the development of a variety of feed supplements. Most research on dietary supplements for pigs has focussed on the very stressful period of life around weaning. At this point the gut is vulnerable to many weaning stressors, especially the abrupt changes occurring in microbiota composition and loss of sow milk with its protective properties. More recent research has emphasized targeting nutritional interventions at the prenatal and early postnatal phases of growth due to the importance of the neonatal phase for gut development and programming (Schokker et al., 2015, 2018).

http://dx.doi.org/10.19103/AS.2021.0089.09

Monitoring the developing gut microbiota provides the best way to assess postnatal maturation of the mucosal immune system and barrier function (Schokker et al., 2015, 2018; Le Bourgot et al., 2017), affecting health and performance later in life (Zwirwitz et al., 2019; Ruczizka et al., 2019). Research has shown the ability of fermentable carbohydrates to stabilize and control the postnatal gut microbial environment. Fermentable oligosaccharides have been included in diets for farm animals since the mid-1980s in Japan, mainly to improve weight gain and prevent diarrhoea (Hidaka et al., 1985, 1986). Since then, prebiotics have been evaluated for their ability to selectively stimulate beneficial gut bacteria, mainly lactobacilli and bifidobacteria, or for their ability to support the establishment of probiotic cultures in the porcine gut (Trevisi et al., 2008). The advent of in-depth sequencing techniques, especially the broader accessibility of 16S rRNA amplicon sequencing, has revolutionized our view of the major bacterial taxa in the GIT of pigs and thus our understanding of diet-microbe and microbe-to-microbe/pathogen interactions (Bindels et al., 2015).

This chapter summarizes research on the effects of prebiotics on the porcine GIT. It discusses the definition, mechanisms and application of prebiotic oligosaccharides in promoting gut microbiota and function. Given the growing body of evidence on the importance of prenatal and early postnatal nutrition on intestinal development, the chapter focusses particularly on the modulatory capacity of prebiotics in promoting gut health at weaning during the early neonatal phase of development. Since current research emphasizes the significance of porcine milk oligosaccharides (PMO) in the development of commensal microbiota, as compared to short-chain fructo- (FOS) and galacto-oligosaccharides (GOS) and lactose present in sow milk, the chapter also discusses PMO in more detail.

2 Maintenance of gut health and functionality

A plethora of intestinal components contributes to gut health. These include the gut microbiota (including the presence/absence of pathogens) and the immune system, as well as digestive, absorptive and secretory functions along the GIT. Intestinal factors, such as nutrient digestion and absorption, host metabolism and energy generation, microbiota load and diversity, chemical and physical barriers, and mucosal immune responses, all contribute to gut health (Kogut, 2019). Prebiotics help modulate the complex microbial community in the gut in contributing to the health and well-being of the host.

Commensal gut bacteria contribute to gut health in various ways, for example, by controlling pathobionts, programming gut development and functioning as well as delivery of nutrients (e.g. vitamins and fermentation metabolites). Gut bacteria are not only commensals but maintain a symbiotic relationship with their host (Markowiak and Slizewska, 2017). The intestinal immune system needs to learn to respond appropriately to the residing microbiota (i.e. overt activation vs. tolerance) (Frosali et al., 2015; Broom and Kogut, 2018; Guevarra et al., 2019). A healthy gut also affects system homeostasis, reflected in the host's ability to withstand environmental and infectious stressors (Kogut, 2019).

Any intervention early in life can promote or disrupt the development and maintenance of a stable commensal gut microbiota, with consequences for gut function and performance later in life (Schokker et al., 2015; Guevarra et al., 2019; Ruczizka et al., 2019). When this balance is disrupted, for example, by dysbiosis, pathogens may take the opportunity to multiply and activate a (local) immune response. Abrupt dietary changes (e.g. weaning) can also lead to rapid activation of the intestinal immune response and loss of immune tolerance. Changes in diet cause dramatic shifts in the microbial community. They expose the host to a different subset of microbiota-associated molecular patterns (MAMP), structural motifs that are highly conserved in microbes present on outer bacterial cells. MAMPs are recognized by specialized pattern recognition receptors (PRR) in the intestinal mucosa. This leads to the activation of cellular pro-inflammatory pathways via key adaptor proteins (e.g. Myeloid differentiation primary response 88 (MyD88)) and transcription factors (e.g. nuclear factor kappa-light-chain-enhancer of activated B cells (NF-κB)). This also leads to production of various cytokines and chemokines, which signal to immune (including adaptive) cells in the underlying gut-associated lymphoid tissue (GALT) (Abreu, 2010; Broom and Kogut, 2018).

3 The porcine gut microbiome

The taxonomic composition and metabolic activity of the gut microbiota are shaped by a number of complex internal and external factors (Guevarra et al., 2019). These include differences in inter-region conditions in the GIT, principally related to function, such as available substrates for growth, pH, redox potential, digesta transit time, mucus production and composition. They are also affected by host antimicrobial secretions, including antimicrobial peptides, defensins and immunoglobulin A (IgA), as well as age and inter-individual variation (Guevarra et al., 2019; Kogut, 2019). Intra-region variation in environmental conditions also contributes to differences between luminal and mucosa-associated populations (Kelly et al., 2017). Luminal microbes compete with the host for digestible nutrients along the GIT. They contribute to host digestion by converting otherwise unusable dietary components, such as fermentable carbohydrates, into absorbable nutrients, for example, medium and short-chain fatty acids (SCFA) (Flint et al., 2015). Bacteria also influence each other due to competition for substrates and niches, affecting their gut environment via secondary bacterial metabolites, such as antimicrobial and quorum-sensing molecules (Louis et al., 2007). Metabolic cross-feeding between bacteria also plays an important role in substrate breakdown along the GIT (Louis et al., 2007).

While all regions of a pig's stomach are densely colonized by bacteria (Motta et al., 2017), density is particularly high in the saclike non-glandularis region of the porcine stomach given the specific gastric physiology of pigs. In this region, the digesta pH remains higher and only incomplete mixing of digesta with those from the corpus and fundus regions takes place. This allows the development of a much more diverse microbial community in both the mucosa and digesta (Metzler-Zebeli et al., 2013; Mann et al., 2014; Motta et al., 2017; Newman et al., 2018). Although this part of the stomach is a promising target for the use of short-chain prebiotics in pig nutrition, research has focussed more on prebiotic effects on the microbiota communities in the lower regions of the GIT. This is despite the role that gastric fermentation of lactose and other short-chain carbohydrates present in sow milk plays during the suckling phase, facilitating the postprandial decrease in gastric pH of the neonatal piglet.

In-depth sequencing shows that taxa from all major phyla (i.e. Firmicutes, Bacteroidetes, Proteobacteria, Tenericutes, Fusobacteria, Actinobacteria, Spirochaetes, Fibrobacteres and Synergistetes) are present in gastric digesta. Their abundance depends on the particular gastric niche as well as on the dietary composition (Mann et al., 2014; Motta et al., 2017; Newman et al., 2018). In the small intestine, the bacterial load increases from the proximal to distal regions (10^6 to 10^9 gene copies/g digesta), reflecting the water content of digesta. Similar bacterial taxa can be found as in the stomach but abundance differs (Mann et al., 2014; Kelly et al, 2017; Newman et al., 2018). Since transit time in the small intestines is short (2–4 h), the contribution of bacteria to overall fermentation is low compared to the large intestines. Nevertheless, many short-chain prebiotics, such as FOS, are completely fermented when they reach the terminal ileum (Houdijk et al., 2002). This suggests significant bacterial action in the upper porcine GIT and the need to control this activity. This is particularly important when considering that the small intestine is a colonization site for many porcine enteric pathogens, such as Salmonella, pathogenic *Escherichia coli* strains, *Lawsonia* and *Brachyspira*.

The cecum and proximal sections of the colon are the regions with the highest microbial density and diversity in the porcine GIT, amounting from 10^{11} to 10^{12} gene copies/g digesta (Kelly et al., 2017; Newman et al., 2018; Guevarra et al., 2018). Due to the high fermentability of short-chain oligosaccharides, prebiotic effects may be smaller in the large intestine, especially in older pigs. However, due to microbial dependencies and interactions, indirect effects of these prebiotics may be traceable, affecting bacterial abundance and metabolic activity.

The immediate postnatal period is the most critical phase in the establishment of the gut microbial population. During this initial maturation period, the GIT undergoes significant growth, with the developing gut microbiota influencing morphological and functional differentiation processes and immune development (Schokker et al., 2018). This early colonization of the gut determines the reactivity of the immune system in later phases of life (Alizadeh et al., 2016). A clear understanding of the bacteria making up the porcine commensal gut microbiota is therefore critical in understanding dietary prebiotic effects in this early phase of life.

The gut of the neonatal piglet undergoes a remarkable shift from a more or less germ-free state to an extremely dense and diverse microbial population (Guevarra et al., 2019). Although the concept of strict successional changes during microbial colonization has been challenged (Frese et al., 2015), it is generally agreed that the porcine GIT is first inhabited by early colonizers belonging to Enterobacteriaceae and Clostridiaceae (Ruczizska et al., 2019; Shresta et al., 2020). As facultative anaerobes, *Escherichia* can tolerate the aerobic conditions in the neonatal gut and hence dominate in the first days of life. Enterobacteriaceae lower the redox potential in the neonatal gut, supporting the growth of strict anaerobes which appear during the first week of life (Bezirtzoglou, 1997; Bian et al., 2016; De Rodas et al., 2018). *Clostridium* is also a typical early colonizer of the neonatal gut, often preceding anaerobic bacteria like *Bacteroides* (Bezirtzoglou, 1997). Anaerobic *Fusobacterium* and *Bacteroides* predominate until about 2 weeks of life, whereas Ruminococcaceae, Veillonellaceae, and Christensenellaceae only appear in higher abundances in the later preweaning phase (Frese et al., 2015; Ruczizka et al., 2019; Shresta et al., 2020).

Later developments in microbial composition depend on whether the piglet is reared on sow milk alone or has access to milk replacer, creep feed or other edible organic material in the environment. With the introduction of plant-based feed, bacterial genera that only appear after weaning, such as Prevotellaceae, Ruminococcaceae and Lactobacillaceae, increase in numbers (Bian et al., 2016). The postweaning bacteria outnumber bacterial taxa whose function is to facilitate consumption of milk (Frese et al, 2015). *Prevotella* in particular has been associated with plant polysaccharide consumption; it remains present in high numbers until the establishment of an adult-like microbial community (Guevarra et al., 2019).

4 Definition of prebiotics

Though definitions have changed over time, prebiotics are classically defined as fermentable carbohydrates that 'result in specific changes in the composition and/or activity of the gastrointestinal microbiota when being fermented, thus conferring benefit(s) upon host health' (Gibson et al., 2010). Functional abilities that a fermentable carbohydrate must fulfil include resistance to low gastric pH and hydrolysis by host enzymes in the upper digestive tract. A prebiotic carbohydrate needs to pass into the large intestines to exert beneficial effects by being fermented by the commensal bacteria (Gibson and Roberfroid, 1995). Oligosaccharides, such as FOS and GOS, have been shown to pass the upper GIT of monogastric animals without being hydrolyzed or absorbed, and are then used in the colon as a substrate for beneficial bacteria such as bifidobacteria and lactobacilli, thereby improving host health (Gibson et al., 2010).

However, contradicting the common view that prebiotics are resistant to enzymes in the body, recent results from pigs demonstrate that brush border membrane vesicles from the small intestine possess β-galactosidase activity, providing evidence of existing transglycosylation activity of mammalian small intestinal glycosidases (Lesbia et al., 2019). These findings are supported by in vitro digestion models using rat small intestinal extract, also demonstrating selective digestion of GOS in the rat small intestine (Ferreira-Lazarte et al., 2017a,b).

Bindels et al. (2015) have suggested that advancing the understanding of diet-microbiome-host interactions challenge important aspects of the current concept of prebiotics, especially the requirement for effects to be 'selective' or 'specific'. The concept of prebiotics originates from human nutrition and its definition is specific to the human gut, where the majority of microbial action on food residues occurs in the large intestine (Gibson et al., 2010). In contrast, easily fermentable carbohydrates are broken down by the microbiota in the porcine stomach and small intestines before reaching the large intestines, causing a prebiotic effect in pigs that diverges from that in humans. As an example, depending on the actual chain length, FOS and inulin are partly or completely degraded by the time they reach the distal ileum in weaned pigs (Houdijk et al., 1999; Metzler-Zebeli et al., 2017). Host-specific differences in microbial colonization need to be taken into account when transferring gut health effects of potential prebiotic candidates from humans to pigs or vice versa.

5 Prebiotic di-oligosaccharides in pig nutrition

Prebiotic effects are not easily predictable, depending on the nature of the carbohydrate and host factors including the individual gut microbiota and the composition of the (basal) diet. Although prebiotic carbohydrates are present in natural feed ingredients (Crittenden and Playne, 2008), it is a common strategy to supplement pig diets with purified components to ensure adequate intake. The efficacy of prebiotic oligosaccharides in promoting gut homeostasis depends on several factors. One of them is the amount of prebiotic-acting carbohydrate present in the natural feedstuffs as natural amounts vary. Pigs in conventional rearing systems usually have very limited access to pasture and dietary sources like fruits and vegetables. Their diet typically consists of cereals and protein-rich legumes which are potential sources for various types of prebiotic oligosaccharides (Campbell et al., 1997; Crittenden and Playne, 2008).

Prebiotic oligosaccharides added to pig diets as purified supplements commonly have a degree of polymerization in the range from 3 to 60. These oligosaccharides may have similar or different monosaccharides, various linkage structures, and may have linear or branched chains (Grieshop et al., 2001). They consist mainly of the monosaccharides glucose, fructose, galactose, mannose and xylose (Manning and Gibson, 2004). The fermentation of prebiotics commences in the porcine stomach and is influenced by sugar monomers, chain length, polymerization degree and type of glycosidic bonds between sugar moieties (Swennen et al., 2006). The most consistent benefits in pigs have been obtained using fructan-type oligosaccharides. Recent research has focussed on use of GOS in milk replacers, creep feed and pre-starter diets, mostly due to results from prebiotic research in human infants for which the suckling piglet is used as model (Salcedo et al., 2016).

5.1 Fructans

Short-chain FOS have been reported to be constituents of the milk of mammals including the pigs (Le Bourgot et al., 2017). Fructans, such as inulin and FOS, are also commonly found in edible parts of a variety of plants (Campbell et al., 1997). While fructan-rich plants such as onion, Jerusalem artichoke, chicory, leek and garlic do not play a significant role in pig diets, other plants containing fructans, such as wheat, rye and barley, are common ingredients in pig diets. Fructans are mixtures of fructose polymers differing in linkage type (β-2,1 or β-2,6), degree of polymerization and branching (Avila-Fernandez et al., 2016). Inulin and FOS are characterized by a degree of polymerization from 3 to 9 for FOS and up to 60 for inulin of fructose monomers, linked by β-2,1-glycosidic bonds with a terminal glucose residue from which the fructan chain is elongated (Louis et al., 2007).

Chicory root is the main source of extraction for commercial production of inulin. FOS can be obtained either by enzymatic synthesis from sucrose or by hydrolysis of inulin from natural sources such as the roots of chicory, artichoke, yacon, dahlia or agave (Hidaka et al., 1988; Martins et al., 2019). Enzymatic synthesis involves transfructosylation reactions via fructosyltransferases [β-fructofuranosidase (EC 3.2.1.26) or β-D-fructosyltransferase (EC 2.4.1.9)] which act as biocatalysts (Martins et al., 2019; Zhang et al., 2019). While FOS and inulin are most often included in pig diets as purified supplements, natural amounts of fructans present in diets (Table 1) (mainly based on wheat, barley and rye) may be sufficient to cause a prebiotic effect, reducing the need to supplement these diets. Despite this, pig diets are not commonly tested for their fructan content, although this is important given the large variation in cereal fructan contents due to fluctuations in environmental growth conditions.

5.2 Galacto-oligosaccharides

Galacto-oligosaccharides are composed of a variable number of galactose units ranging from 2 to 10. Like FOS they are obtained either by enzymatic synthesis or by extraction and hydrolysis (Martins et al., 2019). The type of linkage between units varies according to their origin and production method.

Table 1 Fructan content in feedstuffs for pigs

Ingredient	Fructan content (% dry matter)	Reference
Cereals		
Wheat	0.7-2.9	Campbell et al., 1997; Verspreet et al., 2015
Wheat bran	0.4	Campbell et al., 1997
Wheat germ	0.47	Campbell et al., 1997
Wheat middlings	0.51	Campbell et al., 1997
Spelt	0.9-1.3	Verspreet et al., 2015
Triticale	1.5-2.9	Verspreet et al., 2015
Rye	3.6-6.6	Campbell et al., 1997; Verspreet et al., 2015
Barley	Traces - 1.0	Campbell et al., 1997; Verspreet et al., 2015
Oat	Traces - 0.2	Campbell et al., 1997; Verspreet et al., 2015
Corn	0	Campbell et al., 1997; Verspreet et al., 2015
Corn gluten feed	0.01	Campbell et al., 1997
Corn gluten meal	0.03	Campbell et al., 1997
Rice	0	Campbell et al., 1997; Verspreet et al., 2015
Rice bran	0	
Legumes and meals		
Soybean meal	0	Campbell et al., 1997
Canola meal	0	Campbell et al., 1997
Others		
Alfalfa meal	0.22	Campbell et al., 1997
Beet pulp	0.01	Campbell et al., 1997
Oat straw	0.08	Campbell et al., 1997
Seaweed	0	Campbell et al., 1997

Two types of GOS can be distinguished. The first are α-GOS or α-galactosides derived from legumes such as soy, peas and lupins. They are mainly composed of raffinose, stachyose and verbascose, which consist of 1, 2 or 3 units of D-galactose linked via α-1-3 bonds and bound to a terminal sucrose (Navarro et al., 2019).

The second type of GOS is β-GOS or trans-GOS. These are prepared from lactose by the action of β-galactosidases of microbial or yeast origin which have transgalactosylation activity (Difilippo et al., 2016; Martins et al., 2019). Commonly, GOS produced by β-transglycosylation have a glucose monomer at their reducing end to which multiple galactose monomers with different linkages (β-1-2, β-1-3, β-1-4, and β-1-6) are attached. β-GOS resemble oligosaccharides found in the milk of mammals, including human, cow and sow milk. Due to this resemblance, β-GOS have unique properties among prebiotics that allow them to be used as supplement in formulas where they mimic the physiological effect of GOS found in milk (Vera and Illanes, 2016; Illanes et al., 2016). Although β-GOS are not found in common feedstuffs, they may find their way into pig diets in the form of by-products from the dairy industry such as fermented milk products including yoghurt and whey (Vénica et al., 2016).

5.3 Milk oligosaccharides

The microbe-regulating capacity of N-glycans in milk for suckling piglets has recently attracted attention (Salcedo et al., 2016; Zhang et al., 2018; Mu et al., 2019). Most research on milk oligosaccharides derives from nutrition research for human infants, for which the suckling piglet constitutes an ideal gut model to simulate human digestive physiology due to the large similarity between both species (Salcedo et al., 2016). After lactose and lipids, oligosaccharides are the third most abundant component in the milk of mammals (Zivkovic et al., 2011; Zhang et al., 2018). However, their significance in gut development is still underestimated in farm animal production. Normally, porcine milk constitutes the main source of the piglet's diet for the first weeks of life. It is associated with high abundances of *Bacteroides, Oscillibacter, Escherichia, Alistipes, Lactobacillus* and *Ruminococcaceae* in faecal microbiomes during the suckling phase (Frese et al., 2015; Mach et al., 2015; Salcedo et al., 2016; Mu et al., 2019). Bacterial community composition has also been shown to be relatively stable throughout lactation when piglets were exclusively fed porcine milk (Frese et al., 2015).

Lactose and milk oligosaccharides modulate the neonatal microbiota by acting as prebiotics (Tao et al., 2010). Since it is digestible at the brush border in the small intestine, disaccharide lactose acts as a prebiotic in the stomach of the suckling piglet. Lactose concentrations are generally lower in porcine colostrum with 3.5 g/L and increase up to 5.1 g/L in mature milk (Theil et al., 2014). Lactose and milk oligosaccharides are fermented to lactic and acetic acid in the stomach, essential for the postprandial drop in gastric pH, while shaping the development of the gastric and intestinal commensal microbiota in suckling piglets (e.g. Frese et al., 2015; Guevarra et al., 2019). *Bifidobacterium* requires α-1, 2-fucosylated oligosaccharides for growth (Hao et al., 2019), which may explain why bifidobacteria in pigs are found in higher abundances in the suckling phase compared to postweaning, and why external short-chain FOS showed 'bifidogenic' properties in studies with suckling and weaned pigs (e.g. Shen et al., 2010; Schokker et al., 2018).

Milk oligosaccharides are typically composed of 10 repeating monosaccharide units, including glucose (Glc), galactose (Gal), *N*-acetyl-glucosamine (GlcNAc), *N*-acetyl-galactosamine (GalNAc), fucose (Fuc) and sialic acids (NeuAc/NeuGc; Salcedo et al., 2016). The core unit of milk oligosaccharides can be either lactose [Gal(β-1, 4)Glc] or *N*-acetyl-lactosamine [Gal(β -1, 4)GlcNAc; Aldredge et al., 2013]. Based on their chemical composition, oligosaccharides are either categorized as acidic (containing sialic acid, NeuAc/NeuGc) or neutral (containing GlcNAc, GalNAc or fucose, lacking sialic acid; Salcedo et al., 2016). Similar oligosaccharides are also found in various secretions including saliva, tears and (intestinal) mucus (Hao et al., 2019). Species-specific gut microbial colonization occurs by selectively nourishing genetically compatible bacteria in suckling animals with a complex array of free oligosaccharides (Zivkovic et al., 2011; Salcedo et al., 2016).

A comprehensive overview of oligosaccharides present in the milk of cows, goats, sheep, pigs, horses and dromedary camels has been published by Albrecht et al. (2014). In general, most data for milk oligosaccharide composition is available for human breast milk, for which more than 250 N-glycans have been described though not all characterized (Wu et al., 2010, 2011; Hao et al., 2019). It has been estimated that the amount of oligosaccharides in human milk is 20–1000-fold higher than in bovine milk and with a far greater variety (Hao et al., 2019). Despite the increasing interest in porcine milk N-glycome, information on PMO composition is still limited compared to data for human and bovine milk. The available data suggests that porcine milk has lower oligosaccharide diversity than human milk and is closer to the oligosaccharide diversity reported for bovine and caprine milk (Salcedo et al., 2016). Around 22-33 different oligosaccharides have so far been reported for porcine milk (Tao et al., 2010; Albrecht et al., 2014; Salcedo et al., 2016; Wei et al., 2018; Mu et al., 2019). Up to 90 different PMOs have been identified (Zhang et al., 2018). Breed-specific differences (Mu et al., 2019), together with analytical discrepancies, may explain different reported numbers of PMOs. When comparing the N-glycans in porcine milk with bovine and human milk, porcine milk seems to fall in the middle between human and bovine milk with regards to its fucosylation pattern but appears to be lower in sialylated oligosaccharides (Salcedo et al., 2016; Mu et al., 2019).

Reports on fucosylation and sialylation patterns of PMOs are inconsistent, contributing to the large differences in the PMO profiles between studies (Tao et al., 2010; Albrecht et al., 2014; Salcedo et al., 2016; Mu et al., 2019). To give an example, Mu et al. (2019) found a total of 22 oligosaccharides in milk from Yorkshire and Meishan sows using hydrophilic interaction ultra-performance liquid chromatography, of which 8 were fucosylated (36%), 9 sialylated (41%) and 3 mannosylated (14%). This study showed that breed and lactation stage influenced the relative abundances of PMOs. The relative abundance of fucosylated PMO (GlcNAc4-Man3-Fuc) and sialylated PMO (GlcNAc4-Man3-Gal2-NeuAc) was found to increase during lactation. In addition, Meishan sow milk comprised more mannosylated PMO (GlcNAc2-Man6) and sialylated PMO (GlcNAc5-Man3-Gal-NeuAc) than Yorkshire sow milk. In contrast, Salcedo et al. (2016) reported a relative increase in fucosylated PMOs but a decrease in sialylated PMOs with progressing lactation in Yorkshire and Hampshire sows. As the authors stated, the sample preparation procedure and analytical technique used often hinder comparisons of PMO profiles among studies (Salcedo et al., 2016).

PMO content is generally highest in colostrum on day 1 of lactation and then decreases during early lactation (Tao et al., 2010). Differences in the abundance of milk oligosaccharides among individual dams may also play a role as has been reported for human mothers (Hao et al., 2019). The ability to secrete fucosyl α-1, 2 oligosaccharides in milk was missing in some human mothers, helping to explain variation in the developing gut microbiota in infants (Hao et al., 2019). This difference may also help to explain the variation in piglets' gut microbiome composition between litters (Mu et al., 2019). In human mothers, the lack of fucosylated oligosaccharides has been associated with lower abundance and diversity of bifidobacteria in their infants, causing poor health and obesity later in life (Zivkovic et al., 2011; Hao et al., 2019).

6 Modes of action of prebiotics

The basic principle behind manipulation of the gut microbiota using prebiotics is that certain species benefit when their preferred substrate is available in greater abundance (Flint et al., 2015). The genome determines the outcome of bacterial competition for substrates of host and dietary origin. Gut microbiota generally compete most for carbohydrates, especially in the hindgut, with depletion of fermentable carbohydrates from digestion and fermentation (Metzler-Zebeli et al., 2019). Prebiotic oligosaccharides should selectively stimulate bacterial taxa that are (preferably) characterized by saccharolytic fermentation. As the traditional targets of prebiotics, enzymes responsible for oligosaccharide degradation have been more thoroughly investigated in probiotic cultures including *Lactobacillus* and *Bifidobacterium* strains (e.g. Rossi et al., 2005; Mano et al., 2018).

A range of β-fructofuranosidases have been identified in the breakdown of fructans. Bacterial β-fructofuranosidases vary in their ability to cleave β-2,1 and β-2,6 bonds in sucrose, FOS, inulin and levan-type fructans, depending on the fructan source and bacterial strains including *Bacteroides*, *Lactobacillus* and *Bifidobacterium*, *Roseburia* and *Eubacterium* (as reviewed by Louis et al., 2007; Mano et al., 2018). Many different *Bifidobacterium* species can utilize short-chain FOS (Pokusaeva et al., 2011), whereas only few *Bifidobacterium* species are capable of using long-chain FOS, depending on the initiation of degradation by other gut bacteria (Biedrzycka and Bielecka, 2004; Meyer and Stasse-Wolthuis, 2009). Similarly, the *lacZ* gene coding for β-galactosidase, necessary for bacterial GOS catabolism, can be found in a variety of bacteria present in the porcine GIT, including *Bacteroides*, *Prevotella*, *Parabacteroides*, *Clostridium*, *Bifidobacterium* and *Eubacterium* (Metzler-Zebeli et al., 2019). *Christensenellaceae* also seem to thrive on GOS and dairy-based diets (Azcarate-Peril et al., 2017).

The effect of prebiotic oligosaccharides on the porcine GIT combines changes in microbial metabolite profiles together with host immune and mucosal metabolic response (Fig. 1). Though not further discussed in this chapter, commensals promoted by dietary prebiotics produce a vast array of

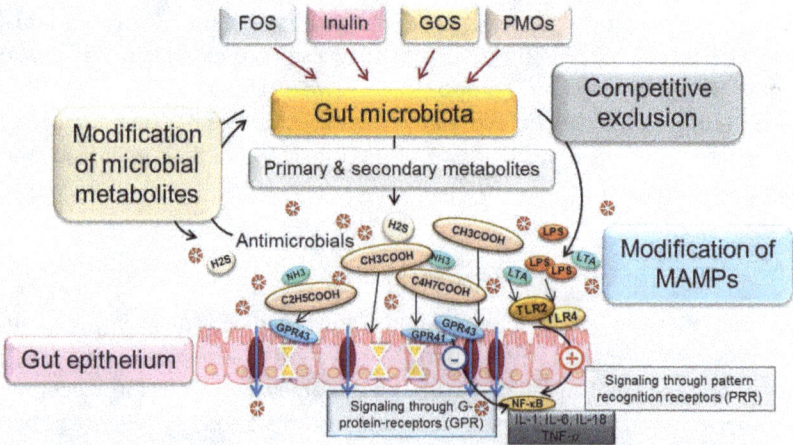

Figure 1 Modes of action of prebiotic oligosaccharides. FOS, fructo-oligosaccharides; GOS, galacto-oligosaccharides; IL, interleukin; LPS, lipopolysaccharides; LTA, lipoteichoic acid; MAMP, microbiota-associated molecular pattern; PMO, porcine milk oligosaccha-rides; TLR, toll-like receptor; TNF-α, tumour-necrosis-factor-α.

potentially bioactive compounds, including antimicrobial peptides, thereby suppressing the growth of opportunistic pathogens (Broom and Kogut, 2018). The modes of action of primary fermentation metabolites (i.e. SCFA) are discussed below. Together with changes in fermentation profiles, alterations in bacterial numbers are accompanied by changes in MAMPs (e.g. bacterial flagellin, peptidoglycan and LPS). LPS, which is abundant in the outer cell membrane of gram-negative bacteria, in particular acts as strong immune stimulant, upregulating a variety of innate immune responses while weakening the mucosal barrier function (Broom and Kogut, 2018).

Alongside gases the primary metabolites of saccharolytic fermentation are SCFA (Flint et al., 2015), with formate, acetate, propionate and butyrate as the major products (Morrison and Preston, 2016). Concentrations of SCFAs in the porcine gut vary, depending on microbiota activity. Concentrations are controlled by transit time, pH and nutrient concentrations in the various regions of the gut, ranging from 5 mM to 150 mM (Metzler-Zebeli et al., 2013; Newman et al., 2018). Medium-chain fatty acids, such as lactate and succinate, are also produced from saccharolytic fermentation. However, they are quickly metabolized into acetate, propionate, butyrate and caproate via microbial cross-feeding relationships, especially in the hindgut (Flint et al., 2015). Higher concentrations of lactate can be found in the porcine stomach and small intestines. Relatively minor amounts of branched-chain fatty acids and valerate are also produced mainly through fermentation of protein-derived branched chain amino acids. Intestinal concentrations increase with progressive depletion of carbohydrates in the hindgut regions (Newman et al., 2018).

Acetate production pathways are widely distributed among bacterial groups across all major phyla found in the porcine GIT, whereas pathways for propionate, butyrate and lactate production appear to be more substrate specific (Morrison and Preston, 2016). Lactate is a major bacterial fermentation product in suckling piglets and is produced by bifidobacteria as well as by lactic acid bacteria in the *Lactobacillales* order, including the genera *Lactobacillus*, *Enterococcus*, *Pediococcus*, *Lactococcus* and *Weissella* (Morrison and Preston, 2016).

The link between diet, microbiome composition, SCFA production and the host, although relatively well characterized, is still difficult to predict due to many influencing factors and methodological limitations (Morrison and Preston, 2016). Before being absorbed, SCFA significantly affect the gut environment, influencing pH, gut motility, nutrient uptake, microbial balance and fermentative activity (Louis et al., 2007). The SCFA are absorbed via diffusion and carrier-mediated transport along the GIT and are used by the enterocytes as an energy source (McKenzie et al., 2017). Enterocytes use SCFA as respiratory fuel in a preferential order with butyrate being the most favoured substrate. Other

SCFA, such as valerate, caproate, propionate and acetate, contribute equally to ATP generation in the enterocytes (Jørgensen et al., 1997).

There has been recent focus on the role of SCFA as modulators of gut physiology and development, including the development of the immune system (McKenzie et al., 2017). Research has highlighted their anti-inflammatory and anti-apoptotic effects (mainly butyrate and propionate) on gut mucosa (McKenzie et al., 2017; Shimizu et al., 2019). The SCFA bind to G-protein coupled receptors (GPR), such as GPR43, GPR41 and GPR109A. These receptors play fundamental roles in the regulation of inflammatory responses, influencing epithelial integrity, as well as IgA antibody responses and macrophage, regulatory T-cell and dendritic cell activities (McKenzie et al., 2017; Sun et al., 2017). In addition, SCFAs exert effects via manipulation of gene transcription, specifically through inhibition of histone deacetylase expression or function. Treatment of macrophages with butyrate has been shown to down-regulate LPS-induced pro-inflammatory mediators, including nitric oxide, interleukin (IL)-6 and IL-12 (Chang et al., 2014). These effects were independent of toll-like receptor (TLR) and GPR signalling, but seemed to be due to the inhibition of histone deacetylases by butyrate. SCFA effects are not limited to the gut lumen. After absorption, SCFAs exert systemic effects through circulating in the plasma, influencing neural activity, the inflammatory response, glucose homeostasis, and lipid metabolism, endocrine hormone and gut hormone secretion (Shimizu et al., 2019).

In the intestinal lumen, the production of adequate and balanced SCFAs by a healthy gut microbiota is an important factor that inhibits colonization by common foodborne pathogens and controls virulence factor expression in enteric pathobionts (Nakanishi et al., 2009; Jacobson et al., 2018). The same principle is behind the supplementation of weaned pig diets with organic acids (e.g. formic acid, fumaric acid, lactic acid or sorbic acid) (Metzler et al., 2005). The antimicrobial effect of SCFA is related to a decrease in intracellular pH and a specific anion effect on cell metabolism and replication in susceptible bacteria (Metzler et al., 2005). In adequate concentrations, SCFAs have direct antimicrobial activity against pathogenic bacteria. *Bacteroides* spp. have been shown to mediate resistance to *Salmonella* colonization by propionate production (Jacobson et al., 2018). Propionate decreases the expression of *Salmonella* genes located in Salmonella Pathogenicity Island 1 (Gantois et al., 2006). The protective and antimicrobial effects of SCFAs are concentration dependent. Dysregulation of intestinal SCFA production seems to facilitate intestinal colonization by pathogens (Lamas et al., 2019). If intestinal SCFA levels fall below a certain threshold, the inhibitory effect of SCFA such as butyrate on the expression of virulence genes ceases, as shown, for instance, for enterohemorrhagic *Escherichia coli* and *Campylobacter jejuni* (Luethy et al., 2017).

In general, intestinal SCFA levels decline due to weaning. Dysregulated SCFA production may contribute to dysbiosis in the early-weaned pig in addition to the loss of SCFA-induced inhibition of pathogens. SCFA, particularly butyrate, are degraded in the enterocytes via β-oxidation. When intestinal butyrate levels are low due to depletion of butyrate-producing commensals, the enterocyte switches to anaerobic ATP-generation from lactate and consumes less oxygen (Byndloss et al., 2017; Gillis et al., 2018; Stecher and Jung, 2018). Simultaneously, epithelial cells upregulate nitrate generation due to the missing butyrate. Increased abundance of oxygen and nitrate may foster the growth of aero-tolerant anaerobes at the mucosa, especially *Enterobacteriaceae*. *Enterobacteriaceae* including taxa like *Escherichia coli* and *Salmonella* have been shown to utilize L-(+)-lactate in vitro, giving them another growth advantage (Stecher and Jung, 2018). Although only reported in mice so far, similar effects may occur in the GIT of newly weaned pigs. Aside from the dietary provision of sufficient amounts of fermentable carbohydrates for the intestinal production of SCFA, the other critical factor is that the piglet continues to eat after separation from the sow at weaning (Bauer et al., 2011).

7 Optimization of gut function by fructans and galacto-oligosaccharides (GOS)

Supplementing weaner diets with FOS, inulin and GOS has been investigated as a useful strategy to counteract susceptibility to infection associated with the stress of weaning (Flickinger et al., 2003). Until the advent of in-depth sequencing, targeted approaches focussing on a limited number of bacterial taxa were often used to study prebiotic effects on the gut microbiota. This meant that the full impact of prebiotic oligosaccharides on bacterial taxonomy was not assessed. In addition, results from studies on the effects of FOS and GOS on gut microbial taxonomic composition and activity have been inconsistent due to differences in experimental methods. These differences include the type and dosage of the prebiotic used, the composition of the basal diet, age of the pig, the time over which the diet was given, the gut region of interest as well as the methodology used to target the bacterial community (Houdijk et al., 2002; Tanner et al., 2014; Metzler-Zebeli et al., 2017).

High proportions of fructan-rich wheat and barley included in the basal diet may mask potential prebiotic effects (Houdijk et al., 2002; Metzler-Zebeli et al., 2009). Supplementation levels tested in diets for suckling and weaned pigs have ranged, for instance, from 0.15% to 10% for FOS (Tsukahara et al., 2003; Le Bourgot et al., 2017) and from 0.25% to 4% for GOS (Houdijk et al., 2002; Alizadeh et al., 2016; Bouwhuis et al., 2017). Dietary inulin levels ranged from 0.15% to 25.6% (Metzler-Zebeli et al., 2017). As a result of these differences, a number of studies of inulin have demonstrated beneficial modulation of the gut microbiota, improved growth and gut health (Konstantinov et al., 2004; Loh et al., 2006; Patterson et al., 2010), whereas others failed to find a similar effect in postweaning pigs (McCormack et al., 2019).

In some studies, different types of fructans and β-GOS were investigated together, allowing direct comparison of two different types of oligosaccharides on gut microbiota and function in the same experimental setting. Patterson et al. (2010) compared short-chain and long-chain inulin and a 50:50 mixture of both (all added at 4% to the diet) on intestinal bacteria in weaned pigs using terminal-restriction-fragment-length polymorphism and found increased numbers of bifidobacteria and lactobacilli at the jejunal mucosa and content, whereas *Clostridium* and members of *Enterobacteriaceae* were depressed along the GIT. Mikkelsen and Jensen (2004) and Mountzouris et al. (2006) did not detect an increase of bifidobacteria or lactobacilli after 1% or 4% FOS inclusion in the diet nor after 1% or 4% β-GOS inclusion. Using fluorescence-in-situ-hybridization, Tzortzis et al. (2005) found higher numbers of bifidobacteria in the colonic content of weaned pigs fed diets containing 4% β-GOS, whereas 1.6% inulin proved insufficient to increase their numbers. Both prebiotics had no effect on levels of *Clostridium histolyticum* and *Bacteroides* (Tzortzis et al., 2005). Tanner et al. (2014) simulated the porcine proximal colon using the PolyFermS in vitro continuous fermentation model and 454-pyrosequencing. They found different effects of 8 g/d of FOS and GOS on *Bifidobacteriaceae*, *Enterobacteriaceae* and *Prevotellaceae* while levels of *Lachnospiraceae* increased. As *Lachnospiraceae* comprise a number of butyrate-producing bacteria, they have contributed to increased butyrate production (Tanner et al., 2014).

Gut microbial composition and SCFA can directly impact the colonization ability of enteric pathogens such as *Salmonella* (Pieper et al., 2009). Using the same PolyFermS in vitro continuous fermentation model, Tanner et al. (2014) demonstrated the capacity of 8 g/d FOS and GOS to inhibit *Salmonella Typhimurium* from becoming established, which the authors attributed to the significant increase in SCFA production from use of prebiotics, especially acetate and propionate. In contrast Bouwhuis et al. (2017) did not find a growth-inhibiting effect of 0.25% GOS on *Salmonella Typhimurium* in growing pigs. They found that GOS supplementation increased intestinal lactobacilli numbers and reduced the expression of pro-inflammatory cytokines (i.e. *IL6*, *IL22*, *TNFA* and *REG2G*) in pig colonic tissue.

A recent meta-analysis confirmed an inhibiting effect of inulin (mean dietary level of 3%–4%) on colonic and faecal enterobacteria including *Escherichia coli* (Metzler-Zebeli et al., 2017). However, contrary to earlier studies (Van Loo, 2004), it also found that inulin reduced colonic bifidobacteria and faecal lactobacilli numbers, with stronger effects in younger pigs which decreased with age (Metzler-Zebeli et al., 2017). Nevertheless, dietary inulin may help to lower stomach pH in weaned pigs, which may be favourable for gut health in this stressful period of life (Metzler-Zebeli et al., 2017).

Studies report mixed effects of fructan on intestinal fermentation. Some authors have reported modulatory action of long-chain fructans (inulin) on hindgut fermentation profiles in finisher pigs, though with no effect in reducing boar taint (Salmon and Edwards, 2015). About 4% FOS (but not β-GOS) (Mikkelsen and Jensen, 2004) and 3% inulin (Loh et al., 2006) increased the molar proportion of butyrate in cecal and proximal colonic content in weaned pigs. However, Tanner et al. (2015) found no effect of FOS (8 g/day) on cecal and colonic SCFA in adult Göttingen minipigs. The meta-analysis by Metzler-Zebeli et al. (2017) suggested only small effects of inulin on primary fermentation metabolites, showing a positive trend of ileal lactic acid concentrations with increasing dietary inulin levels.

Reflecting results for microbiota effects of FOS and GOS, results for gut mucosal response are similarly mixed, depending largely on the mucosal parameters of interest (e.g. digestive and absorptive functions versus mucosal barrier and immune functions), pig age and experimental design (e.g. dietary composition, inclusion level, time on diets). Some authors reported that dietary supplementation with short-chain FOS and GOS increased intestinal villus height, crypt depth, mitotic cells and the number of mucin-producing goblet cells in both neonatal and weaned piglets (Tsukahara et al., 2003; Alizadeh et al., 2016; Le Bourgot et al., 2017; Schokker et al., 2018). These effects were partly associated with increased intestinal butyrate concentrations, especially in the colon (Tsukahara et al., 2003). However, other studies found that FOS at dietary inclusion levels of, for instance, 0.25% and 3.0% had no effect on disaccharidase activities and villus height in the small intestine when fed to weaned pigs for 3 weeks (Shim et al., 2005b). Some studies observed immune-stimulating effects of FOS and GOS, showing an enhanced immune responsiveness of the intestinal mucosa postweaning (Le Bourgot et al., 2017).

Changes in nutrient metabolism, potentially leading to better metabolic capability, were reported for pigs that received 2% inulin in their diet throughout the postweaning and fattening period (McCormack et al., 2019). There was higher expression of genes involved in glucose absorption and homeostasis (i.e. *SGLT1*, *GIP* and *GLP1*) in the duodenal mucosa of inulin-supplemented pigs, whereas these pigs had lower maltase activity at the duodenal brush border compared to their control counterparts (McCormack et al., 2019). In contrast, inulin supplementation had no effect on gut structure, including villus height, villus width, crypt depth, the villus height:crypt depth ratio and goblet cells at slaughter age. Interestingly, serum cholesterol and white blood cell and granulocyte counts decreased due to inulin supplementation. The lipid modulatory action of inulin may have been associated with increased intestinal acetate production, whereas lower counts of immune cells may have been linked to the lower relative abundance of potential pathogens (*Campylobacter* and *Chlamydia*) and higher abundance of lactic acid bacteria observed in the faeces and digesta of these pigs (McCormack et al., 2019).

8 Prebiotic effects on gut functions in the early postnatal phase

Intervention with short-chain prebiotics may be beneficial in the (early) neonatal period, a critical time for intestinal priming, including the development of digestive, barrier and immune functions and the first build-up of an immune tolerance towards commensal microbiota (Gomez de Aguero et al., 2016; Schokker et al., 2015, 2018). Experimental approaches used to study the prebiotic capabilities of oligosaccharides in the early neonatal phase are partly based on studies for human infants. As for the postweaning and fattening periods, prebiotic effects during the postnatal period vary. In addition, due to the series of changes in the developing gut microbiome, bacterial responses to oligosaccharides differ with increasing age of the piglet. This was demonstrated, for instance, in a human-microbiota-associated pig model using PCR-denaturing gradient gel electrophoresis (Shen et al., 2010). This investigated the effects of short-chain FOS (0.5 g/kg body weight/day from day 1 to 37 of life) on 'non-bifidobacteria', including the *Clostridium leptum* subgroup, *Bacteroides*, *Subdoligranulum variabile* and *Faecalibacterium*-related species. The study found these fluctuated, depending on the developmental stage of the suckling and weaned piglet. In contrast, FOS promoted bifidobacteria continuously throughout the neonatal and postweaning period (Shen et al., 2010).

A growing body of evidence emphasizes the impact that the gut microbial composition of the mother sow has on bacterial community development in the neonatal GIT of her offspring. This is an important factor for the stabilization of the gut homeostasis in suckling piglets as preparation for weaning (Paßlack et al., 2015). The effect of supplementing the diet of gestating and lactating sows with FOS and inulin on the diet-sow-piglet axis has been the focus of several studies (e.g. Paßlack et al., 2015; Le Bourgot et al., 2014, 2017). Supplementation of the gestation and lactation diet of sows with 3% inulin raised the faecal abundance of enterococci in sows and the cecal content of their offspring. It also increased the *Clostridium leptum* cluster in piglets born from inulin-fed sows on day 10 of life (Paßlack et al., 2015). Although the effects of inulin on bacterial taxa in sows and offspring were generally small, they showed the importance of the mother-piglet relationship for microbial development. These observations were supported by those of Le Bourgot et al. (2017) who reported higher microbial metabolite concentrations in faeces during the suckling period (total faecal SCFA (P<0.05)) and in colonic content after weaning (28 days postweaning; butyrate (P<0.10)) in piglets born from sows receiving 10 g of short-chain FOS daily during the last third of gestation and throughout lactation.

While using the mother sow to transmit prebiotic effects on her offspring is probably easier to apply in practice, another approach to programme the gut in the early neonatal phase is to administrate piglets with prebiotics. This enables a more direct and controllable prebiotic effect on gut maturation and immune development in the neonate than indirectly via the gut microbiota of the sow (Schokker et al., 2018; Wang et al., 2019). Schokker et al. (2018) administrated FOS (5 g/day) to suckling piglets from day 2 to 14 of life, and then compared the intestinal development of suckling piglets on day 14 and 25 of life. In jejunal digesta, there was no 'bifidogenic' effect of FOS as observed in the colon of the supplemented piglets. In contrast, piglets receiving the FOS had more *Escherichia coli, Turicibacter* but less *Lactobacillus* species in the jejunum on day 14 of life (Schokker et al., 2018). This effect was not only obvious at the end of the supplementation period, but continued until day 25 of life, indicating a lasting effect on microbial composition in the suckling phase. In colonic digesta, short-chain FOS stimulated lactobacilli and bifidobacteria along with *Lachnospiraceae* including *Roseburia* and *Megasphaera, Eubacteriaceae, Pasteurellaceae* (i.e. *Actinobacillus*) and *Bacteroidia* (Schokker et al., 2018). Since this study investigated only the FOS effects before weaning, no conclusion can be drawn on whether effects on the gut microbiome affected gut homeostasis in the critical first days of the postweaning period. Similar modulatory effects of supplementing milk replacers with GOS or oral administration in neonatal piglets have also been reported (Table 2).

Although there have only been a limited number of studies, research demonstrates certain immune-modulating effects of fructans and GOS in neonatal piglets (Le Bourgot et al., 2017; Schokker et al., 2018; Wang et al., 2019). Prebiotic consumption by gestating and lactating sows may therefore be an important route to modulate their offspring's intestinal immunity. Coinciding with the changes in bacterial growth and metabolism in the gut, maternal fructan supplementation may impact immune signalling in their offspring (Schokker et al., 2018). However, due to the small number of available studies, more research under different experimental and applied conditions is needed to confirm the immunomodulatory effect of maternally administrated fructans and GOS in neonatal piglets. It also needs to be established whether prebiotic-induced reinforcement of the gut mucosal barrier in the suckling phase is a short-term effect, relying on the continuing supply of maternal bacteria and microbial metabolites, or whether this effect can be maintained until after weaning, when the bacterial supply from the mother dam has stopped. Some evidence exists from the study of Le Bourgot et al. (2017) where maternal supplementation with 10 g of short-chain FOS appeared to strengthen both physical non-specific intestinal defences and mucosal immune response in the

Table 2 Potential effects of galactooligosaccharides (GOS) on gut microbiota and function in neonatal and growing pigs

Study	Breed	Age/length experimental period	GOS level	Administration form	Gut microbiota	Gut function	Other effects
Alizadeh et al., 2016	Landrace × Yorkshire	Neonatal, day 3-27 of life	0.80%	Milk replacer	Day 12: faecal *Bacteroides*↑; day 26: faecal lactobacilli, bifidobacteria ↑, cecal pH ↓, cecal butyrate ↑	Day 3: duodenum: villus height, mucosa area ↑; jejunum, ileum, cecum: *CLDN1, ZO1, ZO2* expression ↑; colon: *OCLN, ZO1, ZO2* expression ↑; β-defensin 2 expression ↑; day 26: duodenum, jejunum: villus height, width ↑, *OCLN, ZO1, ZO2* expression ↑; jejunum: lactase, sucrase, maltase ↑; cecum, colon: *CLDN1* expression ↑	Days 19, 22, 26: saliva immunoglobulin A ↑
Difilippo et al., 2015	Landrace × Yorkshire	Neonatal, day 2-28 of life	0.80%	Milk replacer	-	Disappearance of GOS in gut: 99.9% (0.1% of intake in faeces)	-
Wang et al., 2019	Duroc × Landrace × Yorkshire	Neonatal, day 1-21 of life	1 g GOS/ kg BW	Oral administration day 1-8 of life	Day 8: colon: total SCFA, acetate, butyrate ↑; *Fusobacterium, Dorea* ↓, *Prevotella* ↑ day 21: colon: propionate ↑; *Fusobacterium* ↓, *Parabacteroides*, *Clostridium XI* ↑	Day 8: colon: *IL10, TFG1B* expression ↑; Day 21: colon: *IL8, IL10, MUC1, MUC2, MUC4* expression ↑	-
Bouwhuis et al., 2017	-	Growing, 30.9 kg, experimental period 17 days	0.25%	Wheat-barley-soybean meal diet	Cecum, colon: *Lactobacillus* ↑	Colon: *IL6* ↓, *IL18* ↑, *IL22* ↓, *TNFA* ↓, *REG3G* ↓ expression	

offspring postweaning, supporting the concept of nutritional programming of the immune system with prebiotics. However, gut samples were only collected 16 days postweaning, missing the critical period of the immediate postweaning phase. Nevertheless, maternal short-chain FOS increased the cecal number of mucin-secreting goblet cells together with a higher ileal cytokine production (interferon-γ, interleukin-4, and tumour-necrosis-factor-α) in the weaned pigs (Le Bourgot et al., 2017). Moreover, pigs from sows fed the FOS diet had a 75%-enhanced specific IgA response in serum to *Lawsonia intracellularis* vaccine challenge (Le Bourgot et al., 2017).

An effective host immune response to foodborne and enteric pathogens is critical for resilience to disease (Kogut, 2017), and especially needed in the early postweaning period when maternal IgA are no longer available to the piglet. Effects of FOS that were passed from the sow to the offspring were different to those when short-chain FOS were only provided in the postweaning diet, indicating that the indirect prebiotic effect via the sow was through alterations in microbial taxonomy and activity in sow's faeces or milk (Le Bourgot et al., 2017). This evidence is supported by an earlier study of the same group (Le Bourgot et al., 2014). This showed that supplementation of the maternal diet during gestation and lactation with short-chain FOS stimulated the maturation of the neonatal intestinal immune system by improving the passive immunization through increasing colostral concentrations of IgA and immunosuppressive transforming growth factor (TGF)-β1. Maternal short-chain FOS supplementation led to a greater density of cells in mesenteric lymph node and Peyer's patch cells and stimulated their secretory activity towards more T helper cell type 1 cytokines (interferon (INF)-γ) and higher levels of secretory IgA in 21-day-old piglets (Le Bourgot et al., 2014).

In addition to boosting the immune system, research has explored the effects of short-chain oligosaccharides on intestinal structure and functioning in suckling piglets. When administrating a FOS solution orally from day 2 to 14 of life, Schokker et al. (2018) found greater jejunal villus height and crypt depth in 25-day-old piglets and differentially expressed gene sets in the jejunal mucosa at 14 and 25 days of life, whereas (immune) gene expression in the colonic mucosa (despite changes in the bacterial activity) was unaffected by FOS. At 14 days of life, genes related to the cell cycle in the jejunal mucosa were less expressed, whereas the expression of genes for extracellular matrix processes was upregulated in the jejunum of FOS-supplemented piglets. At 25 days of age, FOS-supplemented piglets expressed fewer immune genes in the jejunal mucosa compared to control piglets (Schokker et al., 2018).

9 Gut effects of porcine milk oligosaccharides

Research on the influence of PMOs on the gut has so far focussed primarily on gut microbiota development in suckling piglets. Piglets drinking only sow milk developed a faecal bacterial metagenome enriched by enzymes active on milk-derived glycans, which are otherwise indigestible to the host animal (Frese et al., 2015; Salcedo et al., 2016). PMOs help shape the gut microbiota into a 'milk-oriented microbiome' characterized by elevated numbers of *Bacteroides, Oscillibacter, Escherichia, Lactobacillus* and *Ruminococcus,* thereby contributing to neonatal health (Frese et al., 2015). *Bacteroides* have been particularly associated with milk-glycan utilization (especially sialic acid) during the suckling period (Frese et al., 2015; Bian et al., 2016; De Rodas et al., 2018). Primary metabolites from PMOs produced during in vitro fermentation with faecal inoculum from 21-day-old piglets included acetate (49%), propionate (27%), butyrate (20%), lactate (2%) and succinate (1%; Difilippo et al., 2016), reflecting the broad use of PMOs by the various gut bacteria.

Postweaning, this sialic acid-consuming microbiota disappears, as PMOs are replaced by dietary starch (Frese et al., 2015; Salcedo et al., 2016). *Prevotella,* which is associated with plant cell-wall degradation, replaces *Bacteroides* populations. Enterobacteriaceae, which comprised more than 53% of total assigned metagenomic reads pre-weaning, decreased to only 0.6% after weaning (Salcedo et al., 2016). Enterobacteriaceae encode genes related to the consumption of free fucose but lack the enzymatic capacity to liberate these monomers from milk glycans. The loss of PMOs offers a competitive advantage for potential pathogens in the ecosystem after weaning (Ng et al., 2013; Salcedo et al., 2016). Retaining bacterial taxa in the gut associated with a 'milk-oriented microbiome' may be an effective strategy in the prevention of enteric systemic disorders typically associated with weaning such as oedema disease and diarrhoea. Aside from the stimulation of commensal bacteria, milk oligosaccharides prevent pathogen binding to the gut epithelium, with fucosylated and sialylated milk glycans acting as binding targets for the pathogen (Newburg et al., 2005; Bode and Jantscher-Krenn, 2012; Yu et al., 2013; Charbonneau et al., 2016). More specifically, fucosylated and sialylated milk oligosaccharides share common structural motifs with glycans on the intestinal epithelia known to be receptors for pathogens, implying a defensive strategy to protect the neonate from disease (Newburg et al., 2005; Yu et al., 2013).

While the importance of PMOs on the gut bacterial maturation has been studied to some extent in piglets, little information is available on their impact on enterocyte and immune system development via gut microbiota stabilization. Certain anti-inflammatory effects, effects on gut barrier function and stimulation of enterocytes by milk oligosaccharides have been reported (Boudry et al., 2017; Salcedo et al., 2016; Bering, 2018). However, the effectiveness of milk oligosaccharides seemed to be linked to a certain maturation stage of the gut, since BMO did not improve intestinal permeability and prevent enterocolitis in a preterm pig model (Bering, 2018).

Most studies of the modulatory effect of milk oligosaccharides on gut functions has been based on rodent models or preterm pig models using BMO or human milk oligosaccharides (Boudry et al., 2017; Bering, 2018). Due to the species-characteristic differences in milk oligosaccharide composition, these models are far away from representing the normal sow-piglet-relationship and the situation on farms in practice. As the composition of PMOs appear to change over the course of lactation (Mu et al., 2019), their potentially different roles at the various stages of gut maturation need to be taken into consideration.

The effects of milk oligosaccharides in newborns are not limited to the gut and the gut microbiota modulation as these oligosaccharides are also essential for neuronal development (Ten Bruggencate et al., 2014). Although the majority of milk oligosaccharides are fermented by the gut microbiota, small amounts of milk oligosaccharides cross the intestinal epithelium intact, reaching the systemic circulation, whilst another small fraction is excreted in faeces (Difilippo et al., 2015). To support intestinal absorption, in vitro studies indicate that neutral and acidic milk oligosaccharides pass the intestinal lining transcellularly through the enterocytes (Gnoth et al., 2001; Ten Bruggencate et al., 2014). Most studies are based on human milk oligosaccharides and BMOs after ingestion of human breast milk and infant formula (Goehring et al., 2014; Ruhaak et al., 2014), with little information on intestinal absorption of PMOs from sow milk and milk replacers. The systemic appearance of BMO in serum and urinary excretion has been reported for piglets that received a milk replacer based on cow's milk. The BMO were sialylated and neutral oligosaccharides, including 3- and 6-sialyllactose, disialyllactose, and N-acetylgalactosaminyl lactose (Difilippo et al., 2015). Since some PMO fractions resemble those of BMOs, it may be assumed that intestinal absorption may be similar. However, due to factors such as species-specific profiles, amounts of milk oligosaccharides and potential effects of 'species-nonspecific' oligosaccharides on gut barrier function, these findings should be interpreted with care. The specific role of PMOs on gut development, barrier and immune functions in suckling piglets still needs to be clarified in order to supplement milk replacers' age appropriate with PMOs.

10 Future trends in research

β-GOS and FOS are used in functional foods for adults and extensively employed in infant formula to stimulate the development of the newborn microbiota (Martins et al., 2019). Manufacturers of human infant formula have started to fortify them with commercial GOS, polydextrose and FOS in an attempt to more closely mimic human milk (Hao et al., 2019). As suckling piglets in big litters are provided with milk replacer to ensure sufficient energy and nutrient intake for growth and development from the first days of life, using β-GOS and FOS as prebiotic carbohydrates may shape the beneficial commensal gut microbiota during the preweaning stage. In addition, continuous supplementation of GOS and FOS in creep feed, pre-starter and starter diets may enable the persistence of commensal taxa beyond the suckling phase, thereby reducing the immune-stimulating shifts in the gut microbial composition and loss of immune-tolerance towards the newly establishing plant-oriented gut microbiota.

Most commercial milk replacers aimed for piglets are still mainly formulated on the basis of the macronutrient content of porcine milk, comprising bovine whey as source for lactose, hydrolyzed starch, soy protein and (plant-based) medium-chain saturated fatty acid-rich fat sources, such as palm oil or coconut oil. To date, few porcine milk replacers and creep feed diets contain FOS and GOS, but often contain combinations of other supplements to boost gut health and maximize performance. Though bovine whey contains a portfolio of milk oligosaccharides, BMOs differ from PMOs in their amount, diversity and structural composition. It is justified to question whether BMOs can support the development of a resilient commensal gut microbiota in piglets. Although other (bioactive) components in the milk replacer than milk oligosaccharides contribute to this effect, gut microbiome research comparing milk replacer and bovine milk with sow milk shows that the source of milk clearly determines microbial community composition and maturation (Poulsen et al., 2017). There is strong evidence of species-specific milk glycan composition as a driving force for gut microbial colonization (Zivkovic et al., 2011; Salcedo et al., 2016) and addressing continuing gut health issues after weaning. Mimicking milk glycan composition in future porcine milk replacer formulations may be an additional nutritional support to avert gut dysbiosis and failure to thrive postweaning. Especially for big litters, where supplementation with milk replacers is indispensable right after birth, addition of prebiotic oligosaccharides, including age-appropriate levels and composition of short-chain FOS and β-GOS as well as of PMOs, may ensure normal microbial colonization and development in the neonatal porcine gut to promote host immune function. The fortification of creep feed and prestarter diets in the future with chemically synthesized PMOs may offer dual functionality, slowing down the dietary change-related (and still often abrupt) shift in the gut microbial composition, thereby maintaining host immune tolerance and keeping virulence gene expression in pathobionts in check, while inhibiting mucosal colonization with pathogens. This basic nutritional strategy of selectively but continuously nourishing gut bacteria with a complex array of oligosaccharides, including short-chain FOS, GOS and PMOs needs to be explored more thoroughly in future-feeding protocols aimed at suckling and weaned piglets.

11 Where to look for further information

The following articles provide a good overview on animal and human milk oligosaccharides:

- Albrecht, S., Lane, J. A., Marino, K., Al Busadah, K. A., Carrington, S. D., Hickey, R. M. and Rudd, P. M. (2014). A comparative study of free oligosaccharides in the milk of domestic animals. *Br. J. Nutr.* 111, 1313–1328. doi:10.1017/S0007114513003772.
- Chichlowski, M., German, J. B., Lebrilla, C. B. and Mills, D. A. (2011). The influence of milk oligosaccharides on microbiota of infants: opportunities for formulas. *Annu. Rev. Food Sci. Technol.* 2, 331–351. doi:10.1146/annurev-food-022510-133743.
- Jeong, K., Nguyen, V. and Kim, J. (2012). Human milk oligosaccharides: the novel modulator of intestinal microbiota. *BMB Rep.* 45(8), 433–441. doi:10.5483/BMBRep.2012.45.8.168

The following articles provide a good overview on the biotechnological production of fructan-type oligosaccharides and GOS:

- Flores-Maltos, D. A., Mussatto, S. I., Contreras-Esquivel, J. C., Rodríguez-Herrera, R., Teixeira, J. A. and Aguilar, C. N. (2016). Biotechnological production and application of fructooligosaccharides. *Crit. Rev. Biotechnol.* 36(2), 259–267. doi:10.3109/07388551.2014.953443.
- Mano, M. C. R., Neri-Numa, I. A., da Silva, J. B., Paulino, B. N., Pessoa, M. G. and Pastore, G. M. (2018). Oligosaccharide biotechnology: an approach of prebiotic revolution on the industry. *Appl. Microbiol. Biotechnol.* 102(1), 17–37. doi:10.1007/s00253-017-8564-2.

12 References

Abreu, M. T. (2010). Toll-like receptor signalling in the intestinal epithelium: how bacterial recognition shapes intestinal function. *Nat. Rev. Immunol.* 10(2), 131–144.

Albrecht, S., Lane, J. A., Marino, K., Al Busadah, K. A., Carrington, S. D., Hickey, R. M. and Rudd, P. M. (2014). A comparative study of free oligosaccharides in the milk of domestic animals. *Br. J. Nutr.* 111(7), 1313–1328. doi:10.1017/S0007114513003772.

Aldredge, D. L., Geronimo, M. R., Hua, S., Nwosu, C. C., Lebrilla, C. B. and Barile, D. (2013). Annotation and structural elucidation of bovine milk oligosaccharides and determination of novel fucosylated structures. *Glycobiology* 23(6), 664–676. doi:10.1093/glycob/cwt007.

Alizadeh, A., Akbari, P., Difilippo, E., Schols, H. A., Ulfman, L. H., Schoterman, M. H. C., Garssen, J., Fink-Gremmels, J. and Braber, S. (2016). The piglet as a model for studying dietary components in infant diets: effects of galacto-oligosaccharides on intestinal functions. *Br. J. Nutr.* 115(4), 605–618. doi:10.1017/S0007114515004997.

Ávila-Fernández, Á., Cuevas-Juárez, E., Rodríguez-Alegría, M. E., Olvera, C. and López-Munguía, A. (2016). Functional characterization of a novel β-fructofuranosidase from Bifidobacterium longum subsp. infantis ATCC 15697 on structurally diverse fructans. J. Appl. Microbiol. 121(1), 263–276. doi:10.1111/jam.13154.

Azcarate-Peril, M. A., Ritter, A. J., Savaiano, D., Monteagudo-Mera, A., Anderson, C., Magness, S. T. and Klaenhammer, T. R. (2017). Impact of short-chain galactooligosaccharides on the gut microbiome of lactose-intolerant individuals. Proc. Natl. Acad. Sci. U. S. A. 114(3), E367–E375.

Bauer, E., Metzler-Zebeli, B. U., Verstegen, M. W. and Mosenthin, R. (2011). Intestinal gene expression in pigs: effects of reduced feed intake during weaning and potential impact of dietary components. Nutr. Res. Rev. 24(2), 155–175. doi:10.1017/S0954422411000047.

Bering, S. B. (2018). Human milk oligosaccharides to prevent gut dysfunction and necrotizing enterocolitis in preterm neonates. Nutrients 10(10), 1461. doi:10.3390/nu10101461.

Bezirtzoglou, E. (1997). The intestinal microflora during the first weeks of life. Anaerobe 3(2-3), 173–177.

Bian, G., Ma, S., Zhu, Z., Su, Y., Zoetendal, E. G., Mackie, R., Liu, J., Mu, C., Huang, R., Smidt, H. and Zhu, W. (2016). Age, introduction of solid feed and weaning are more important determinants of gut bacterial succession in piglets than breed and nursing mother as revealed by a reciprocal cross-fostering model. Environ. Microbiol. 18(5), 1566–1577. doi:10.1111/1462-2920.13272.

Biedrzycka, E. and Bielecka, M. (2004). Prebiotic effectiveness of fructans of different degrees of polymerization. Trends Food Sci. Technol. 15(3-4), 170–175. doi:10.1016/j.tifs.2003.09.014.

Bindels, L. B., Delzenne, N. M., Cani, P. D. and Walter, J. (2015). Towards a more comprehensive concept for prebiotics. Nat. Rev. Gastroenterol. Hepatol. 12(5), 303–310. doi:10.1038/nrgastro.2015.47.

Bode, L. and Jantscher-Krenn, E. (2012). Structure-function relationships of human milk oligosaccharides. Adv. Nutr. 3(3), 383S–391S. doi:10.3945/an.111.001404.

Boudry, G., Hamilton, M. K., Chichlowski, M., Wickramasinghe, S., Barile, D., Kalanetra, K. M., Mills, D. A. and Raybould, H. E. (2017). Bovine milk oligosaccharides decrease gut permeability and improve inflammation and microbial dysbiosis in diet-induced obese mice. J. Dairy Sci. 100(4), 2471–2481. doi:10.3168/jds.2016-11890.

Bouwhuis, M. A., McDonnell, M. J., Sweeney, T., Mukhopadhya, A., O'Shea, C. J.and O'Doherty, J. V. (2017). Seaweed extracts and galacto-oligosaccharides improve intestinal health in pigs following Salmonella Typhimurium challenge. Animal 11(9), 1488–1496. doi:10.1017/S1751731117000118.

Broom, L. J. and Kogut, M. H. (2018). Gut immunity: its development and reasons and opportunities for modulation in monogastric production animals. Health Res. 19(1), 46–52. doi:10.1017/S1466252318000026.

Byndloss, M. X., Olsan, E. E., Rivera-Chávez, F., Tiffany, C. R., Cevallos, S. A., Lokken, K. L., Torres, T. P., Byndloss, A. J., Faber, F., Gao, Y., Litvak, Y., Lopez, C. A., Xu, G., Napoli, E., Giulivi, C., Tsolis, R. M., Revzin, A., Lebrilla, C. B. and Bäumler, A. J. (2017). Microbiota-activated PPAR-g signaling inhibits dysbiotic Enterobacteriaceae expansion. Science 357(6351), 570–575. doi:10.1126/science.aam9949.

Campbell, J. M., Bauer, L. L., Fahey, G. C., Hogarth, A. J. C. L., Wolf, B. W. and Hunter, D. E. (1997). Selected fructooligosaccharide (1-Kestose, nystose, and

1F-β-Fructofuranosylnystose) composition of foods and feeds. *J. Agric. Food Chem.* 45(8), 3076-3082. doi:10.1021/jf970087g.

Chang, P. V., Hao, L., Offermanns, S. and Medzhitov, R. (2014). The microbial metabolite butyrate regulates intestinal macrophage function via histone deacetylase inhibition. *Proc. Natl. Acad. Sci. U. S. A.* 111(6), 2247-2252. doi:10.1073/pnas.1322269111.

Charbonneau, M. R., O'Donnell, D., Blanton, L. V., Totten, S. M., Davis, J. C. C., Barratt, M. J., Cheng, J., Guruge, J., Talcott, M., Bain, J. R., Muehlbauer, M. J., Ilkayeva, O., Wu, C., Struckmeyer, T., Barile, D., Mangani, C., Jorgensen, J., Fan, Y. M., Maleta, K., Dewey, K. G., Ashorn, P., Newgard, C. B., Lebrilla, C., Mills, D. A. and Gordon, J. I. (2016). Sialylated milk glycans promote growth in gnotobiotic mice and pigs with a stunted Malawian infant gut microbiota. *Cell* 164(5), 859-871. doi:10.1016/j.cell.2016.01.024.

Crittenden, R. and Playne, M. J. (2008). Nutrition news. Facts and functions of prebiotics, probiotics and synbiotics. In: Lee, Y. K. and Salminen, S. (Eds), *Handbook of Probiotics and Prebiotics.* Wiley-Interscience, Kansas State University, Hoboken, NJ, pp. 535-582.

De Rodas, B., Youmans, B. P., Danzeisen, J. L., Tran, H. and Johnson, T. J. (2018). Microbiome profiling of commercial pigs from farrow to finish. *J. Anim. Sci.* 96(5), 1778-1794. doi:10.1093/jas/sky109.

Difilippo, E., Bettonvil, M., Willems, R. H., Braber, S., Fink-Gremmels, J., Jeurink, P. V., Schoterman, M. H., Gruppen, H. and Schols, H. A. (2015). Oligosaccharides in urine, blood, and feces of piglets fed milk replacer containing Galacto-Oligosaccharides. *J. Agric. Food. Chem.* 63(50), 10862-10872. doi:10.1021/acs.jafc.5b04449.

Difilippo, E., Pan, F., Logtenberg, M., Willems, R. H., Braber, S., Fink-Gremmels, J., Schols, H. A. and Gruppen, H. (2016). Milk oligosaccharide variation in sow milk and milk oligosaccharide fermentation in piglet intestine. *J. Agric. Food Chem.* 64(10), 2087-2093. doi:10.1021/acs.jafc.6b00497.

Ferreira-Lazarte, A., Olano, A., Villamiel, M. and Moreno, F. J. (2017a). Assessment of in vitro digestibility of dietary carbohydrates using rat small intestinal extract. *J. Agric. Food Chem.* 65(36), 8046-8053. doi:10.1021/acs.jafc.7b01809.

Ferreira-Lazarte, A., Montilla, A., Mulet-Cabero, A.-I., Rigby, N., Olano, A., Mackie, A. and Vollamiel, M. (2017b). Study on the digestion of milk with prebiotic carbohydrates in a simulated gastrointestinal model. *J. Funct. Foods* 33, 149-154, doi:10.1016/j.jff.2017.03.031.

Flickinger, E. A., Van Loo, J. and Fahey, G. C. Jr. (2003). Nutritional responses to the presence of inulin and oligofructose in the diets of domesticated animals: a review. *Crit. Rev. Food Sci. Nutr.* 43(1), 19-60.

Flint, H. J., Duncan, S. H., Scott, K. P. and Louis, P. (2015). Links between diet, gut microbiota composition and gut metabolism. *Proc. Nutr. Soc.* 74(1), 13-22. doi:10.1017/S0029665114001463.

Frese, S. A., Parker, K., Calvert, C. C. and Mills, D. A. (2015). Diet shapes the gut microbiome of pigs during nursing and weaning. *Microbiome* 3, 28. doi:10.1186/s40168-015-0091-8.

Frosali, S., Pagliari, D., Gambassi, G., Landolfi, R., Pandolfi, F. and Cianci, R. (2015). How the intricate interaction among toll-like receptors, microbiota, and intestinal immunity can influence gastrointestinal pathology. *J. Immunol. Res.* 2015, 489821. doi:10.1155/2015/489821.

Gantois, I., Ducatelle, R., Pasmans, F., Haesebrouck, F., Hautefort, I., Thompson, A., Hinton, J. C. and Van Immerseel, F. (2006). Butyrate specifically down-regulates salmonella pathogenicity island 1 gene expression. *Appl. Environ. Microbiol.* 72(1), 946–949. doi:10.1128/AEM.72.1.946-949.2006.

Gibson, G. R. and Roberfroid, M. B. (1995). Dietary modulation of the human colonic microbiota: introducing the concept of prebiotics. *J. Nutr.* 125(6), 1401–1412.

Gibson, G. R., Scott, K. P., Rastall, R. A., Tuohy, K. M., Hotchkiss, A., Dubert-Ferrandon, A., Gareau, M., Murphy, E. F., Saulnier, D., Loh, G., Macfarlane, S., Delzenne, N., Ringel, Y., Kozianowski, G., Dickmann, R., Lenoir-Wijnkoop, I., Walker, C. and Buddington, R. (2010). Dietary prebiotics: current status and new definition. *Food Sci. Tech. Bull. Funct. Foods* 7(1), 1–19.

Gillis, C. C., Hughes, E. R., Spiga, L., Winter, M. G., Zhu, W., Furtado de Carvalho, T., Chanin, R. B., Behrendt, C. L., Hooper, L. V., Santos, R. L. and Winter, S. E. (2018). Dysbiosis-associated change in host metabolism generates lactate to support Salmonella growth. *Cell Host Microbe.* 23(1), 54–64.e6. doi:10.1016/j.chom.2017.11.006.

Gnoth, M. J., Rudloff, S., Kunz, C. and Kinne, R. K. (2001). Investigations of the in vitro transport of human milk oligosaccharides by a Caco-2 monolayer using a novel high performance liquid chromatography mass spectrometry technique. *J. Biol. Chem.* 276(37), 34363–34370.

Goehring, K. C., Kennedy, A. D., Prieto, P. A. and Buck, R. H. (2014). Direct evidence for the presence of human milk oligosaccharides in the circulation of breastfed infants. *PLoS ONE* 9(7), e101692. doi:10.1371/journal.pone.0101692.

Gomez de Agüero, M., Ganal-Vonarburg, S. C., Fuhrer, T., Rupp, S., Uchimura, Y., Li, H., Steinert, A., Heikenwalder, M., Hapfelmeier, S., Sauer, U., McCoy, K. D. and Macpherson, A. J. (2016). The maternal microbiota drives early postnatal innate immune development. *Science* 351(6279), 1296–1302. doi:10.1126/science.aad2571.

Grieshop, C. M., Reese, D. E. and Fahey, G. C. Jr. (2001). Nonstarch polysaccharides and oligosaccharides in swine nutrition. In: Lewis, A. J. and Southern, L. L. (Eds), *Swine Nutrition* (2nd edn.). Boca Racon, FL: CRC Press.

Guevarra, R. B., Hong, S. H., Cho, J. H., Kim, B. R., Shin, J., Lee, J. H., Kang, B. N., Kim, Y. H., Wattanaphansak, S., Isaacson, R. E., Song, M. and Kim, H. B. (2018). The dynamics of the piglet gut microbiome during the weaning transition in association with health and nutrition. *J. Anim. Sci. Biotechnol.* 9, 54. doi:10.1186/s40104-018-0269-6.

Guevarra, R. B., Lee, J. H., Lee, S. H., Seok, M. J., Kim, D. W., Kang, B. N., Johnson, T. J., Isaacson, R. E. and Kim, H. B. (2019). Piglet gut microbial shifts early in life: causes and effects. *J. Anim. Sci. Biotechnol.* 10, 1. doi:10.1186/s40104-018-0308-3.

Hao, H., Zhu, L. and Faden, H. S. (2019). The milk-based diet of infancy and the gut microbiome. *Gastroenterol. Rep.* 7(4), 246–249 doi:10.1093/gastro/goz031.

Hidaka, H., Eida, T., Hashimoto, K. and Nakazawa, T. (1985). Feeds for domestic animals and methods for breeding them. European Patent Appl. 0133547A2, 15 pp.

Hidaka, H., Eida, T. and Hamaya, T. (1986). Livestock feed containing inulo-oligosaccharides and breeding of livestock by using the same. European Patent Appl. 0171026A2, 11 pp.

Hidaka, H., Hirayama, M. and Sumi, N. (1988). A fructooligosaccharide-producing enzyme from Aspergillus niger ATCC 20611. *Agric. Biol. Chem.* 52, 1181–1187.

Houdijk, J. G., Bosch, M. W., Tamminga, S., Verstegen, M. W., Berenpas, E. B. and Knoop, H. (1999). Apparent ileal and total-tract nutrient digestion by pigs as affected by dietary nondigestible oligosaccharides. *J. Anim. Sci.* 77(1), 148–158. doi:10.2527/1999.771148x.

Houdijk, J. G., Hartemink, R., Verstegen, M. W. and Bosch, M. W. (2002). Effects of dietary non-digestible oligosaccharides on microbial characteristics of ileal chyme and faeces in weaner pigs. *Arch. Tierernahr.* 56(4), 297–307. doi:10.1080/00039420214346.

Illanes, A., Vera, C. and Wilson, L. (2016) Chapter 4.Enzymatic production of galacto-oligosaccharides. In: Illanes, A., Guerrero, C., Vera, C., Wilson, L., Conejeros, R. and Scott, F. (Eds), *Lactose-Derived Prebiotics: A Process Perspective.* Elsevier Academic Press, London, UK. doi:10.1016/B978-0-12-802724-0.00004-4.

Jacobson, A., Lam, L., Rajendram, M., Tamburini, F., Honeycutt, J., Pham, T., Van Treuren, W., Pruss, K., Stabler, S. R., Lugo, K., Bouley, D. M., Vilches-Moure, J. G., Smith, M., Sonnenburg, J. L., Bhatt, A. S., Huang, K. C. and Monack, D. (2018). A gut commensal-produced metabolite mediates colonization resistance to salmonella infection. *Cell Host Microbe.* 24(2), 296–307.e7. doi:10.1016/j.chom.2018.07.002.

Jørgensen, J. R., Clausen, M. R. and Mortensen, P. B. (1997). Oxidation of short and medium chain C2-C8 fatty acids in Spraaue-Dawley rat colonocytes. *Gut* 40(3), 400–405.

Kelly, J., Daly, K., Moran, A. W., Ryan, S., Bravo, D. and Shirazi-Beechey, S. P. (2017). Composition and diversity of mucosa-associated microbiota along the entire length of the pig gastrointestinal tract; dietary influences. *Environ. Microbiol.* 19(4), 1425–1438. doi:10.1111/1462-2920.13619.

Kogut, M. H. (2017). Issues and consequences of using nutrition to modulate the avian immune response. *J. Appl. Poult. Res.* 26(4), 605–612. doi:10.3382/japr/pfx028.

Kogut, M. H. (2019). The effect of microbiome modulation on the intestinal health of poultry. *Anim. Feed Sci. Technol.* 250, 32–40. doi:10.1016/j.anifeedsci.2018.10.008.

Konstantinov, S. R., Awati, A., Smidt, H., Williams, B. A., Akkermans, A. D. and de Vos, W. M. (2004). Specific response of a novel and abundant Lactobacillus amylovorus-like phylotype to dietary prebiotics in the guts of weaning piglets. *Appl. Environ. Microbiol.* 70(7), 3821–3830. doi:10.1128/AEM.70.7.3821-3830.2004.

Lamas, A., Regal, P., Vázquez, B., Cepeda, A. and Franco, C. M. (2019). Short chain fatty acids commonly produced by gut microbiota influence Salmonella enterica motility, biofilm formation, and gene expression. *Antibiotics (Basel)* 8(4), 265. doi:10.3390/antibiotics8040265.

Le Bourgot, C., Ferret-Bernard, S., Le Normand, L., Savary, G., Menendez-Aparicio, E., Blat, S., Appert-Bossard, E., Respondek, F. and Le Huërou-Luron, I. (2014). Maternal short-chain fructooligosaccharide supplementation influences intestinal immune system maturation in piglets. *PLoS ONE* 9(9), e107508. doi:10.1371/journal.pone.0107508.

Le Bourgot, C., Le Normand, L., Formal, M., Respondek, F., Blat, S., Apper, E., Ferret-Bernard, S. and Le Huërou-Luron, I. (2017). Maternal short-chain fructo-oligosaccharide supplementation increases intestinal cytokine secretion, goblet cell number, butyrate concentration and Lawsonia intracellularis humoral vaccine response in weaned pigs. *Br. J. Nutr.* 117(1), 83–92. doi:10.1017/S0007114516004268.

Lesbia, J.-G., Julio-Gonzalez, C., Hernandez-Hernandez, O., Moreno, F. J., Olano, A., Jimeno, M. L. and Corzo, N. (2019). Trans-β-galactosidase activity of pig enzymes embedded in the small intestinal brush border membrane vesicles. 9. doi:10.1038/s41598-018-37582-8.

Loh, G., Eberhard, M., Brunner, R. M., Hennig, U., Kuhla, S., Kleessen, B. and Metges, C. C. (2006). Inulin alters the intestinal microbiota and short-chain fatty acid concentrations in growing pigs regardless of their basal diet. *J. Nutr.* 136(5), 1198–1202. doi:10.1093/jn/136.5.1198.

Louis, P., Scott, K. P., Duncan, S. H. and Flint, H. J. (2007). Understanding the effects of diet on bacterial metabolism in the large intestine. *J. Appl. Microbiol.* 102(5), 1197–1208. doi:10.1111/j.1365-2672.2007.03322.x.

Luethy, P. M., Huynh, S., Ribardo, D. A., Winter, S. E., Parker, C. T. and Hendrixson, D. R. (2017). Microbiota-derived short-chain fatty acids modulate expression of Campylobacter jejuni determinants required for commensalism and virulence. *mBio* 8(3), e00407-17. doi:10.1128/mBio.00407-17.

Mach, N., Berri, M., Estellè, J., Levenez, F., Lemonnier, G., Denis, C., Leplat, J. J., Chevaleyer, C., Billon, Y., Doré, J., Rogel-Gaillard, C. and Lepage, P. (2015). Early-life establishment of the swine gut microbiome and impact on host phenotypes. *Environ. Microbiol. Rep.* 7(3), 554–569. doi:10.1111/1758-2229.12285.

Mann, E., Schmitz-Esser, S., Zebeli, Q., Wagner, M., Ritzmann, M. and Metzler-Zebeli, B. U. (2014). Mucosa-associated bacterial microbiome of the gastrointestinal tract of weaned pigs and dynamics linked to dietary calcium-phosphorus. *PLOS ONE* 9(1), e86950. doi:10.1371/journal.pone.0086950.

Manning, T. S. and Gibson, G. R. (2004). Microbial-gut interactions in health and disease. Prebiotics. *Best. Pract. Res. Clin. Gastroenterol.* 18(2), 287–298.

Mano, M. C. R., Neri-Numa, I. A., da Silva, J. B., Paulino, B. N., Pessoa, M. G. and Pastore, G. M. (2018). Oligosaccharide biotechnology: an approach of prebiotic revolution on the industry. *Appl. Microbiol. Biotechnol.* 102(1), 17–37. doi:10.1007/s00253-017-8564-2.

Markowiak, P. and Slizewska, K. (2017). Effects of probiotics, prebiotics, and Synbiotics on human health. *Nutrients* 9(9), 1021; doi:10.3390/nu9091021.

Martins, G. N., Ureta, M. M., Tymczyszyn, E. E., Castilho, P. C. and Gomez-Zavaglia, A. (2019). Technological aspects of the production of fructo and galacto-oligosaccharides. Enzymatic synthesis and hydrolysis. *Front. Nutr.* 6, 78. doi:10.3389/fnut.2019.00078.

McCormack, U. M., Curião, T., Metzler-Zebeli, B. U., Wilkinson, T., Reyer, H., Crispie, F., Cotter, P. D., Creevey, C. J., Gardiner, G. E. and Lawlor, P. G. (2019). Improvement of feed efficiency in pigs through microbial modulation via fecal microbiota transplantation in sows and dietary supplementation of inulin in offspring. *Appl. Environ. Microbiol.* 85(22), e01255-19. doi:10.1128/AEM.01255-19.

McKenzie, C., Tan, J., Macia, L. and Mackay, C. R. (2017). The nutrition-gut microbiome-physiology axis and allergic diseases. *Immunol. Rev.* 278(1), 277–295. doi:10.1111/imr.12556.

Metzler, B., Bauer, E. and Mosenthin, R. (2005). Microflora Management in the gastrointestinal tract of piglets. *Asian Australas. J. Anim. Sci.* 18(9), 1353–1362.

Metzler-Zebeli, B. U., Ratriyanto, A., Jezierny, D., Sauer, N., Eklund, M. and Mosenthin, R. (2009). Effects of betaine, organic acids and inulin as single feed additives or in combination on bacterial populations in the gastrointestinal tract of weaned pigs. *Arch. Anim. Nutr.* 63(6), 427–441.

Metzler-Zebeli, B. U., Mann, E., Schmitz-Esser, S., Wagner, M., Ritzmann, M. and Zebeli, Q. (2013). Changing dietary calcium-phosphorus level and cereal source selectively alters abundance of bacteria and metabolites in the upper gastrointestinal tracts of weaned pigs. *Appl. Environ. Microbiol.* 79(23), 7264–7272. doi:10.1128/AEM.02691-13.

Metzler-Zebeli, B. U., Canibe, N., Montagne, L., Freire, J., Bosi, P., Prates, J. A. M., Tanghe, S. and Trevisi, P. (2017). Resistant starch reduces large intestinal pH and promotes fecal lactobacilli and bifidobacteria in pigs. *Animal* 13(1), 64–73. doi:10.1017/S1751731118001003.

Metzler-Zebeli, B. U., Newman, M. A., Ladinig, A., Kandler, W., Grüll, D. and Zebeli, Q. (2019). Transglycosylated starch accelerated intestinal transit and enhanced bacterial fermentation in the large intestine using a pig model. *Br. J. Nutr.* 122(1), 1–13. doi:10.1017/S0007114519000849.

Meyer, D. and Stasse-Wolthuis, M. (2009). The bifidogenic effect of inulin and oligofructose and its consequences for gut health. *Eur. J. Clin. Nutr.* 63(11), 1277–1289. doi:10.1038/ejcn.2009.64.

Mikkelsen, L. L. and Jensen, B. B. (2004). Effect of fructo-oligosaccharides and transgalacto-oligosaccharides on microbial populations and microbial activity in the gastrointestinal tract of piglets post-weaning. *Anim. Feed Sci. Technol.* 117(1-2), 107–119.

Morrison, D. J. and Preston, T. (2016). Formation of short chain fatty acids by the gut microbiota and their impact on human metabolism. *Gut Microbes* 7(3), 189–200. doi:10.1080/19490976.2015.1134082.

Motta, V., Trevisi, P., Bertolini, F., Ribani, A., Schiavo, G., Fontanesi, L. and Bosi, P. (2017). Exploring gastric bacterial community in young pigs. *PLoS ONE* 12(3), e0173029. doi:10.1371/journal.pone.0173029.

Mountzouris, K. C., Balaskas, C., Fava, F., Tuohy, K. M., Gibson, G. R. and Fegeros, K. (2006). Profiling of composition and metabolic activities of the colonic microflora of growing pigs fed diets supplemented with prebiotic oligosaccharides. *Anaerobe* 12(4), 178–185. doi:10.1016/j.anaerobe.2006.04.001.

Mu, C., Cai, Z., Bian, G., Du, Y., Ma, S., Su, Y., Liu, L., Voglmeir, J., Huang, R. and Zhu, W. (2019). New insights into porcine milk N-Glycome and the potential relation with offspring gut microbiome. *J. Proteome Res.* 18(3), 1114–1124. doi:10.1021/acs.jproteome.8b00789.

Nakanishi, N., Tashiro, K., Kuhara, S., Hayashi, T., Sugimoto, N. and Tobe, T. (2009). Regulation of virulence by butyrate sensing in enterohaemorrhagic Escherichia coli. *Microbiology* 155(2), 521–530. doi:10.1099/mic.0.023499-0.

Navarro, D. M. D. L., Abelilla, J. J. and Stein, H. H. (2019). Structures and characteristics of carbohydrates in diets fed to pigs: a review. *J. Anim. Sci. Biotechnol.* 10, 39. doi:10.1186/s40104-019-0345-6.

Newburg, D. S., Ruiz-Palacios, G. M. and Morrow, A. L. (2005). Human milk glycans protect infants against enteric pathogens. *Annu. Rev. Nutr.* 25, 37–58. doi:10.1146/annurev.nutr.25.050304.092553.

Newman, M. A., Petri, R. M., Grüll, D., Zebeli, Q. and Metzler-Zebeli, B. U. (2018). Transglycosylated starch modulates the gut microbiome and expression of genes related to lipid synthesis in liver and adipose tissue of pigs. *Front. Microbiol.* 9, 224. doi:10.3389/fmicb.2018.00224.

Ng, K. M., Ferreyra, J. A., Higginbottom, S. K., Lynch, J. B., Kashyap, P. C., Gopinath, S., Naidu, N., Choudhury, B., Weimer, B. C., Monack, D. M. and Sonnenburg, J. L. (2013). Microbiota-liberated host sugars facilitate post-antibiotic expansion of enteric pathogens. *Nature* 502(7469), 96–99. doi:10.1038/nature12503.

Paßlack, N., Vahjen, W. and Zentek, J. (2015). Dietary inulin affects the intestinal microbiota in sows and their suckling piglets. *BMC Vet. Res.* 11, 51. doi:10.1186/s12917-015-0351-7.

Patterson, J. K., Yasuda, K., Welch, R. M., Miller, D. D. and Lei, X. G. (2010). Supplemental dietary inulin of variable chain lengths alters intestinal bacterial populations in young pigs. *J. Nutr.* 140(12), 2158–2161. doi:10.3945/jn.110.130302.

Pieper, R., Bindelle, J., Rossnagel, B., Van Kessel, A. and Leterme, P. (2009). Effect of carbohydrate composition in barley and oat cultivars on microbial ecophysiology and proliferation of Salmonella enterica in an in vitro model of the porcine gastrointestinal tract. *Appl. Environ. Microbiol.* 75(22), 7006–7016.

Pokusaeva, K., Fitzgerald, G. F. and van Sinderen, D. (2011). Carbohydrate metabolism in bifidobacteria. *Genes Nutr.* 6(3), 285–306. doi:10.1007/s12263-010-0206-6.

Poulsen, A. R., de Jonge, N., Sugiharto, S., Nielsen, J. L., Lauridsen, C. and Canibe, N. (2017). The microbial community of the gut differs between piglets fed sow milk, milk replacer or bovine colostrum. *Br. J. Nutr.* 117(7), 964–978. doi:10.1017/S0007114517000216.

Rossi, M., Corradini, C., Amaretti, A., Nicolini, M., Pompei, A., Zanoni, S. and Matteuzzi, D. (2005). Fermentation of fructooligosaccharides and inulin by bifidobacteria: a comparative study of pure and fecal cultures. *Appl. Environ. Microbiol.* 71(10), 6150–6158. doi:10.1128/AEM.71.10.6150-6158.2005.

Ruczizka, U., Metzler-Zebeli, B., Unterweger, C., Mann, E., Schwarz, L., Knecht, C. and Hennig-Pauka, I. (2019). Early parenteral administration of ceftiofur has gender-specific short- and long-term effects on the fecal microbiota and growth in pigs from the suckling to growing phase. *Animals (Basel)* 10(1), 17. doi:10.3390/ani10010017.

Ruhaak, L. R., Stroble, C., Underwood, M. A. and Lebrilla, C. B. (2014). Detection of milk oligosaccharides in plasma of infants. *Anal. Bioanal. Chem.* 406(24), 5775–5784. doi:10.1007/s00216-014-8025-z.

Salcedo, J., Frese, S. A., Mills, D. A. and Barile, D. (2016). Characterization of porcine milk oligosaccharides during early lactation and their relation to the fecal microbiome. *J. Dairy Sci.* 99(10), 7733–7743. doi:10.3168/jds.2016-10966.

Salmon, L. and Edwards, S. A. (2015). The effects of dietary fructo-oligosaccharide addition on boartaint compounds and performance in heavy slaughter weightboars and gilts. *Anim. Feed Sci. Technol.* 207, 130–139. doi:10.1016/j.anifeedsci.2015.05.013.

Schokker, D., Veninga, G., Vastenhouw, S. A., Bossers, A., de Bree, F. M., Kaal-Lansbergen, L. M., Rebel, J. M. and Smits, M. A. (2015). Early life microbial colonization of the gut and intestinal development differ between genetically divergent broiler lines. *BMC Genom.* 16, 418. doi:10.1186/s12864-015-1646-6.

Schokker, D., Fledderus, J., Jansen, R., Vastenhouw, S. A., de Bree, F. M., Smits, M. A. and Jansman, A. A. J. M. (2018). Supplementation of fructooligosaccharides to suckling piglets affects intestinal microbiota colonization and immune development. *J. Anim. Sci.* 96(6), 2139–2153. doi:10.1093/jas/sky110.

Shen, J., Zhang, B., Wei, H., Che, C., Ding, D., Hua, X., Bucheli, P., Wang, L., Li, Y., Pang, X. and Zhao, L. (2010). Assessment of the modulating effects of fructo-oligosaccharides on fecal microbiota using human flora-associated piglets. *Arch. Microbiol.* 192(11), 959–968. doi:10.1007/s00203-010-0628-y.

Shim, S. B., Williams, I. H. and Verstegen, M. W. A. (2005). Effects of dietary fructo-oligosaccharide on villous height and disaccharidase activity of the small intestine, pH, VFA and ammonia concentrations in the large intestine of weaned pigs. *Acta Agric. Scand. A* 55(2–3), 91–97. doi:10.1080/09064700500307201.

Shimizu, H., Masujima, Y., Ushiroda, C., Mizushima, R., Taira, S., Ohue-Kitano, R. and Kimura, I. (2019). Dietary short-chain fatty acid intake improves the hepatic metabolic condition via FFAR3. *Sci. Rep.* 9(1), 16574. doi:10.1038/s41598-019-53242-x.

Shresta, A., Metzler-Zebeli, B. U., Karembe, H., Sperling, D., Koger, S. and Joachim, A. (2020). Shifts in the Fecal Microbial Community of Cystoisospora Suis Infected Piglets in Response to toltrazuril. *Front. Microbiol.* 11, 983. doi:10.3389/fmicb.2020.00983.

Stecher, B. and Jung, K. (2018). LACTATEing salmonella: a host-derived fermentation product fuels pathogen growth. *Cell Host Microbe.* 23(1), 3-4. doi:10.1016/j.chom.2017.12.012.

Sun, M., Wu, W., Liu, Z. and Cong, Y. (2017). Microbiota metabolite short chain fatty acids, GPCR, and inflammatory bowel diseases. *J. Gastroenterol.* 52(1), 1-8. doi:10.1007/s00535-016-1242-9.

Swennen, K., Courtin, C. M. and Delcour, J. A. (2006). Non-digestible oligosaccharides with prebiotic properties. *Crit. Rev. Food Sci. Nutr.* 46(6), 459-471.

Tanner, S. A., Chassard, C., Zihler Berner, A. and Lacroix, C. (2014). Synergistic effects of Bifidobacterium thermophilum RBL67 and selected prebiotics on inhibition of Salmonella colonization in the swine proximal colon PolyFermS model. *Gut Pathog.* 6(1), 44. doi:10.1186/s13099-014-0044-y.

Tanner, S. A., Lacroix, C., Del'Homme, C., Jans, C., Zihler Berner, A., Bernalier-Donadille, A. and Chassard, C. (2015). Effect of Bifidobacterium thermophilum RBL67 and fructo-oligosaccharides on the gut microbiota in Göttingen minipigs. *Br. J. Nutr.* 114(5), 746-755. doi:10.1017/S0007114515002263.

Tao, N., Ochonicky, K. L., German, J. B., Donovan, S. M. and Lebrilla, C. B. (2010). Structural determination and daily variations of porcine milk oligosaccharides. *J. Agric. Food Chem.* 58(8), 4653-4659. doi:10.1021/jf100398u.

Ten Bruggencate, S. J., Bovee-Oudenhoven, I. M., Feitsma, A. L., van Hoffen, E. and Schoterman, M. H. (2014). Functional role and mechanisms of sialyllactose and other sialylated milk oligosaccharides. *Nutr. Rev.* 72(6), 377-389. doi:10.1111/nure.12106.

Theil, P. K., Lauridsen, C. and Quesnel, H. (2014). Neonatal piglet survival: impact of sow nutrition around parturition on fetal glycogen deposition and production and composition of colostrum and transient milk. *Animal* 8(7), 1021-1030. doi:10.1017/S1751731114000950.

Trevisi, P., De Filippi, S., Minieri, L., Mazzoni, M., Modesto, M., Biavati, B. and Bosi, P. (2008). Effect of fructo-oligosaccharides and different doses of Bifidobacterium animalis in a weaning diet on bacterial translocation and toll-like receptor gene expression in pigs. *Nutrition* 24(10), 1023-1029. doi:10.1016/j.nut.2008.04.008.

Tsukahara, T., Iwasaki, Y., Nakayama, K. and Ushida, K. (2003). Stimulation of butyrate production in the large intestine of weaning piglets by dietary fructooligosaccharides and its influence on the histological variables of the large intestinal mucosa. *J. Nutr. Sci. Vitaminol. (Tokyo)* 49(6), 414-421. doi:10.3177/jnsv.49.414.

Tzortzis, G., Goulas, A. K., Gee, J. M. and Gibson, G. R. (2005). A novel galactooligosaccharide mixture increases the bifidobacterial population numbers in a continuous in vitro fermentation system and in the proximal colonic contents of pigs in vivo. *J. Nutr.* 135(7), 1726-1731. doi:10.1093/jn/135.7.1726.

Van Loo, J. (2004). The specificity of the interaction with intestinal bacterial fermentation by prebiotics determines their physiological efficacy. *Nutr. Res. Rev.* 17(1), 89-98. doi:10.1079/NRR200377.

Vénica, C. I., Wolf, I. V., Bergamini, C. V. and Perotti, M. C. (2016). Influence of lactose hydrolysis on galacto-oligosaccharides, lactose, volatile profile and physicochemical parameters of different yogurt varieties. *J. Sci. Food Agric.* 96(15), 4929–4939. doi:10.1002/jsfa.7870.

Vera, C. and Illanes, A. (2016). Chapter 3. Lactose-derived nondigestible oligosaccharides and other high added-value products. In: Illanes, A., Guerrero, C., Vera, C., Wilson, L., Conejeros, R. and Scott, F. (Eds), *Lactose-Derived Prebiotics: A Process Perspective.* Elsevier Academic Press, London, UK. doi:10.1016/B978-0-12-802724-0.00003-2.

Verspreet, J., Dornez, E., van den Ende, W., Delcour, J. A. and Courtin, C. M. (2015). Cereal grain fructans: structure, variability and potential health effects. *Trends Food Sci. Technol.* 43(1), 32–42. doi:10.1016/j.tifs.2015.01.006.

Wang, J., Tian, S., Yu, H., Wang, J. and Zhu, W. (2019). Response of colonic mucosa-associated microbiota composition, mucosal immune homeostasis, and barrier function to early life galactooligosaccharides intervention in suckling piglets. *J. Agric. Food Chem.* 67(2), 578–588. doi:10.1021/acs.jafc.8b05679.

Wei, J., Wang, Z. A., Wang, B., Jahan, M., Wang, Z., Wynn, P. C. and Du, Y. (2018). Characterization of porcine milk oligosaccharides over lactation between primiparous and multiparous female pigs. *Sci. Rep.* 8(1), 4688. doi:10.1038/s41598-018-23025-x.

Wu, S., Tao, N., German, J. B., Grimm, R. and Lebrilla, C. B. (2010). Development of an annotated library of neutral human milk oligosaccharides. *J. Proteome Res.* 9(8), 4138–4151. doi:10.1021/pr100362f.

Wu, S., Grimm, R., German, J. B. and Lebrilla, C. B. (2011). Annotation and structural analysis of sialylated human milk oligosaccharides. *J. Proteome Res.* 10(2), 856–868. doi:10.1021/pr101006u.

Yu, Z. T., Chen, C. and Newburg, D. S. (2013). Utilization of major fucosylated and sialylated human milk oligosaccharides by isolated human gut microbes. *Glycobiology* 23(11), 1281–1292. doi:10.1093/glycob/cwt065.

Zhang, S., Chen, F., Zhang, Y., Yantao Lv, Y., Heng, J., Min, T., Li, L. and Guan, W. (2018). Recent progress of porcine milk components and mammary gland function. *J. Anim. Sci. Biotechnol.* 9, 77. doi:10.1186/s40104-018-0291-8.

Zhang, J., Wang, L., Luan, C., Liu, G., Liu, J. and Zhong, Y. (2019). Establishment of a rapid and effective plate chromogenic assay for screening of Aspergillus species with high β-fructofuranosidase activity for fructooligosaccharides production. *J. Microbiol. Methods* 166, 105740. doi:10.1016/j.mimet.2019.105740.

Zivkovic, A. M., German, J. B., Lebrilla, C. B. and Mills, D. A. (2011). Human milk glycobiome and its impact on the infant gastrointestinal microbiota. *Proc. Natl. Acad. Sci. U.S.A.* 108 (Suppl. 1), 4653–4658. doi:10.1073/pnas.1000083107.

Zwirwitz, B., Pinior, B., Metzler-Zebeli, B., Handler, M., Gense, K., Knecht, C., Ladinig, A., Dzieciol, M., Wetzels, S. U., Wagner, M., Schmitz-Esser, S. and Mann, E. (2019). Microbiota of the gut-lymph node axis: depletion of mucosa-associated segmented filamentous bacteria and enrichment of Methanobrevibacter by colistin sulfate and Linco-Spectin in pigs. *Front. Microbiol.* 10, 599. doi:10.3389/fmicb.2019.00599.

Chapter 4

Recent advances in understanding the role of vitamins in pig nutrition

Charlotte Lauridsen, Aarhus University, Denmark; and J. Jacques Matte, Agriculture and Agri-Food Canada, Canada

1 Introduction

There is, in general, a lack of scientific information on the requirements for vitamins in the modern intensive swine production, as official recommendations are primarily based on genetically outdated lines of pigs and production conditions which have changed dramatically during the last 30 years. Depending upon the source of information, either from Denmark or from Canada, growth and reproduction performance of modern lines of pigs would correspond to daily weight gains over 900 g in the finishing phase and to litter size of 13 (CCSI, 2015) to 16 liveborn/litter (VSP, 2015). Such a level of performance associated with reduced antibiotic usage pig production systems are factors influencing the vitamin requirement, and more information on the optimal vitamin nutrition in this aspect is needed. The level of scientific knowledge for each vitamin is also heterogeneous and empiricism is still present in this field, which is reflected by large variations of vitamin fortifications used in commercial conditions, as reported by previous and recent surveys (BASF, 1993; BASF, 2001; Flohr et al., 2016). Since the middle of the twentieth century, vitamin requirements for pigs have changed from prevention of deficiencies (practically absent nowadays) to optimization of performance for growth and reproduction. However, less information is available on vitamin requirements for optimal immunity and/or meat nutritional quality, although there is a major interest in optimizing

http://dx.doi.org/10.19103/AS.2017.0013.09

these issues through nutritional strategies in the pig production systems. The challenge for the future is to realign objectives from the performance criteria alone to other aspects such as stress and disease resistance.

Besides the decrease in neonatal piglet survival during the last two decades, which has become one of the major problems in modern pig production, pig production had to face several outbreaks of major diseases (e.g. PED, Circovirus and PRRSV). The rapid succession and wide spread of these different infections is a matter of concern both in economic terms and in terms of public good. With this macroscopic view on pig production, it is tempting to hypothesize that modern pig lines are just more fragile and susceptible to pathogens than older pig lines. In parallel, during the last 30 years, changes in modes of production have constantly limited the allowances of major and minor nutrients provided to the pig throughout its life from fertilization of ova to slaughter at market weight. For micronutrients such as vitamins, the development of hyperprolificacy has likely decreased the individual share of their transfer from the mother to embryos/foetus (Matte and Lauridsen, 2013). For fat-soluble vitamins in particular, this might be critical as they are generally poorly transferred to the foetus during gestation (Matte et al., 2014a,b). After birth, the post-natal transfer of fat-soluble vitamins occurs through colostrum and milk, but, given the large litter size with heterogeneous piglets and increased competition between littermates for colostrum (Theil et al., 2014), the partition of vitamins and other micronutrients from the sow to the offspring has also likely been reduced on an individual piglet basis. After weaning, piglets become completely independent from maternal transfer of vitamins and rely on dietary vitamin provision, which is primarily brought by the vitamin premix but may also be provided naturally by ingredients in the diet and/or endogenous synthesis. However, during the critical period after weaning from the sow, piglets may suffer from impaired capability for absorption and digestion, and the dietary composition of the weaning diet may reduce the bioavailability of some vitamins relative to milk.

During the last decades, growth performance has improved constantly towards global reduction of feed intake throughout the entire growing period. Fortification strategies, however, have not actually changed in commercial conditions during at least the last 20 years (BASF, 1993; Flohr et al., 2016). Therefore, it is plausible to assume that daily amounts of vitamins consumed per market pig for the whole starting, growing and finishing period have diminished by the same order of magnitude as feed intake during the last decades. Hence, there is lack of scientific information on the vitamin requirement for the fast-growing lean meat type of pigs according to their physiological responses. Furthermore, during the last decades, some feed regimens have proposed total withdrawal of the vitamin supplementations in premixes during part or for the whole finishing period (McGlone, 2000; Shaw et al., 2002), as an attractive practice to pork producers for reducing feed costs. Although such decrease of vitamin provision to market pigs can be applied without detrimental effect on growth performance (McGlone, 2000), it may have impact on vitamin deposition in carcass and pork products, and subsequently on meat quality of the pork. Despite some discrepancies in the literature probably related in part to experimental design, supplement withdrawal may decrease, for instance, concentrations of pork riboflavin and niacin (Shaw et al., 2002), thiamine and vitamin E (Edmonds and Arentson, 2001), vitamin A (Olivares et al., 2009) as well as vitamin B_{12} and folates (Giguère et al., 2012) . Such decreases in contents of these micronutrients in pork may affect not only meat quality parameters such as oxidative stability, colour and water-binding (Jensen et al., 1998), but also consumer perception for the nutritional value of this meat (Shaw et al., 2002) as pork is generally considered the most important among meat sources consumed by humans. In fact, conventional pork (without withdrawal of the vitamin supplementations) is recognized as an excellent source of several water-soluble vitamins providing a significant percentage (>65% for thiamine and niacin and >25% for riboflavin, vitamin B_6 and vitamin B_{12}) of the recommended daily allowances for humans (Sahlin and House, 2006). Besides the nutritive value of pork, the presence of antioxidative micronutrients might be important to promote oxidative stability. This is especially critical in the context of increasing interest in changing the fatty acid composition with higher proportion of omega-3 which can not only enhance health benefits of this meat for consumers but also challenge its oxidative stability.

Besides impacts on meat quality, another potential consequence of the above-mentioned global reduction of vitamin provisions during the whole life cycle of market pigs may be alteration of immune competence and risk of disease occurrence. Some vitamins play critical roles for this aspect of animal metabolism. This chapter aims to present recent scientific information related to the role of vitamins and their importance for oxidative mechanisms in relation to the development and competence of the immune system which are key contributors to optimal health status of pigs and to their ability to face pathogenic pressure during their life.

2 Supply of vitamins to pigs

2.1 The maternal dependence

In modern pig production, the duration of the post-weaning period until slaughter corresponds roughly to about half of the life duration of a pig (from fertilization of ova to slaughter). This means that for half of their life duration, pigs are entirely dependent on the transfer of vitamins from their mother (*in utero*, colostrum and milk). As mentioned above, this is particularly critical for fat-soluble vitamins. Indeed, vitamins D and E are not being transferred *in utero*, and although local transport of retinoids by retinol-binding protein has been demonstrated in the developing conceptus (Chew et al., 1993), vitamin A is transferred to a limited degree. Thus, as pigs are born almost deficient in vitamins D and E, and with almost no vitamin A depots, they have to rely entirely on the post-natal supply of these vitamins. Colostrum is an important source of especially vitamin E, which is more concentrated (times four) in colostrum than in milk (Lauridsen et al., 2002a). When individual colostrum intake is limited either because of the high number of littermates competing for colostrum or because of a limited synthesis by the sow, this may be a critical factor for vitamin E status in piglets. However, within a short period (four days) after birth, suckling piglets are capable of increasing their plasma vitamin E status by a factor as high as 83 (Lauridsen et al., 2002a). Among livestock, the piglet is born with the lowest vitamin D status, and dietary supplementation of vitamin D to the lactating sow seems not an efficient way of increasing the newborn piglets' vitamin D status (Lauridsen et al., 2010; Matte et al., 2016). Human studies have shown poor penetration of vitamin D and the metabolite 25-hydroxyvitamin D into milk (Kovacs, 2008), whereas, in pigs, no scientific information has been available until recently regarding the vitamin D content in sow colostrum and milk. These latest data indicate that vitamin D metabolites are also present in very small amounts in sow milk with concentrations approximately 10 times smaller than in blood serum (Matte et al., 2017). In fact, in sow milk, two forms, 25-hydroxyvitamin D_3 (25(OH)D_3) and 24,25-dihydroxyvitamin D_3 (24,25(OH)$_2D_3$), are present, the latter accounting roughly for 55–60% of the total detectable vitamin D metabolites (Ouattara et al., unpublished data). Interestingly, although 24,25(OH)$_2D_3$ has long been considered as an inactive form of vitamin D, the literature from poultry (Seo et al., 1997) showed that it may have an important impact on bone metabolism, especially on cartilage formation. The implication of this peculiarity of the sow milk deserves to be further studied. In order to increase piglet survival, major focus has recently been given to the provision of milk supplements either in the form of formulas or in the form of other sources such as bovine colostrum. Use of formula supplemented with vitamin D may be a strategic tool to optimize vitamin D status in piglets, but the bioavailability of other (fat-soluble) vitamins may not be the same in milk formulas as for sow milk. For example, vitamins E and A provided as acetate-bound forms (i.e. α-tocopheryl acetate and retinyl acetate, respectively) should be hydrolysed before absorption. As previously discussed in relation to humans (Lauridsen and Jensen, 2007), premature and low body weight infants may have limited capacity for utilizing acetate-bound vitamin E due to impaired intestinal hydrolysis, and this may also be a focus point in future pig production having a considerable proportion of low birth weight piglets.

In contrast to fat-soluble vitamins, the global perinatal (*in utero* + colostrum) transfer for B vitamins and vitamin C would be favourable to foetuses and piglets (Matte et al., 2014a,b). In fact, it appears that the placental structures allow an active prenatal maternal transfer except for folates and riboflavin. In these last two cases, the maternal transfer is compensated by an important colostral transfer shortly after birth (Matte et al., 2014a,b).

2.2 The post-weaning period

Reduced capacity for absorption of fat-soluble vitamins at weaning and for two-three weeks beyond has also been a focus point in vitamin E research during the last decades. In order to overcome the challenges around weaning with impaired enzyme capacity and lipid absorption, some studies have focused on alternative strategies to vitamin E supplementation, that is, natural versus synthetic vitamin E for sows and piglets (Lauridsen et al., 2002a,b; Amazan et al., 2014). Natural forms may be RRR-α-tocopheryl acetate, or micellized vitamin E (D-α-tocopherol), which, upon supplementation, will increase the concentration of the RRR-α-tocopherol in blood, tissue and cellular membranes of the pig (Lauridsen et al., 2002a,b). Likewise, the metabolite $25(OH)D_3$ has been developed and commercialized (HY-D, DSM Nutritional products) as alternative forms to vitamin D_3 (cholecalciferol) for use in swine nutrition. As shown in Fig. 1, doses greater than 200 IU of $25(OH)D_3$ was more bioavailable than vitamin D_3, and as such, could be considered an equivalent of an even more advantageous source of vitamin D (Fig. 1), especially when pigs are housed indoor without exposure to sunlight. Analyses of the tissue samples (adipose tissue, white and red muscle, and liver) from the same animal experiment (Fig. 1) showed that the content of $25(OH)D_3$ in the different tissues fully correlated with the serum $25(OH)D_3$ level. However, the correlation between the tissue content of vitamin D_3 and serum $25(OH)D_3$ was dependent on the source of ingested vitamin D_3 (Burrild et al., 2016).

In contrast to α-tocopherol, absorbed retinol is re-esterified in the enterocyte and accumulates in the liver. Because of concerns that excess vitamin A intakes in humans are associated with increased risks of hepatotoxicity, bone fracture and teratogenicity, and that elevated dietary vitamin A for pigs reduce vitamin E concentration in pork, withdrawal of vitamin A from pig feed has been studied, (Ayuso et al., 2015). It was concluded that the use of 7.5 times the NRC dietary vitamin A supplementation for long periods was not needed and that short- or long-term vitamin A withdrawal has the potential to reduce feed costs and increase fat and liver α-tocopherol levels without adverse effect on overall growth performance in heavy pigs (Ayuso et al., 2015).

B-vitamins are generally supplemented as synthetic forms from weaning up to slaughter age. Most of those synthetic forms do not present any particular problems in terms of bioavailability, one exception being vitamin B_{12} (Combs, 2012). Cyanocobalamin is the synthetic form of vitamin B_{12} present in

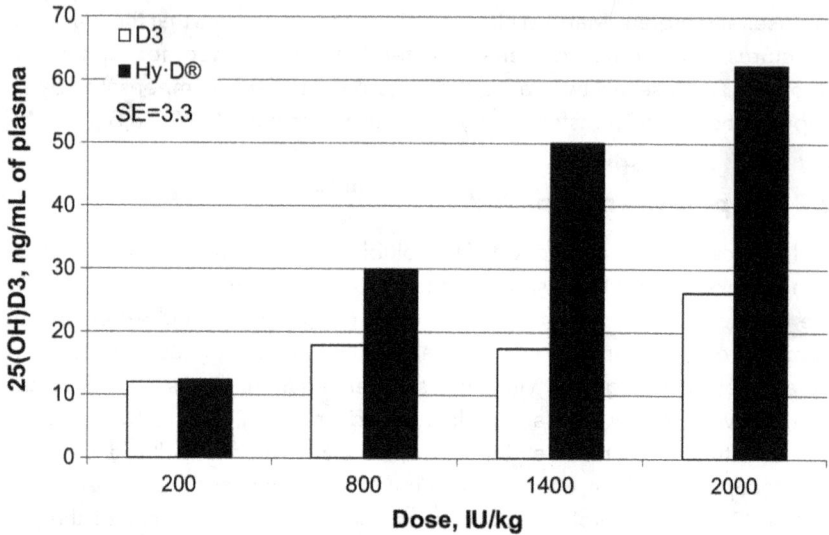

Figure 1 Concentration of 25(OH)D$_3$ (ng/mL) plasma of pigs after feeding an increasing dose of vitamin D in the form of cholecalciferol ('D$_3$') or 25(OH)$_2$D$_3$ (HY·D®). Modified after Lauridsen et al. (2010).

most supplements, the cyanide group being used to stabilize the molecule. However, cyanocobalamin is not biologically active until the cyanide group is enzymatically removed (Herbert, 1988). Bioavailability of the synthetic form of B$_{12}$ is inversely dependent upon the amount given, values being lower than 4% in humans and animals receiving prophylactic or therapeutic levels of supplements (LeGrusse and Watier, 1993; Scott, 1997; Matte et al., 2010). Using pigs as an animal model for humans, Matte et al. (2012) compared the net portal flux of B$_{12}$ (an indicator of intestinal absorption) after ingestion of a natural source (cow milk) versus an equivalent amount of cyanocobalamin and observed that vitamin B$_{12}$ in cow's milk is substantially more bioavailable (values of approximately 10% of the intake) than the synthetic form of this vitamin. Two explanations were raised for interpreting these results, one in relation to the molecular form of this vitamin in cow milk, mostly adenosylcobalamin (Farquharson and Adams, 1976; Fie et al., 1994) and another one, to the presence of specific components in milk, probably casein, facilitating the absorption of this vitamin. Such information could be a basis for designing a more efficient synthetic source of vitamin B$_{12}$. Opportunities for the development of new vitamin B$_{12}$ products are conceivable because actually, in pig production, most of this vitamin (>95%) from synthetic sources is excreted in manure besides the fact that it is one the most expensive vitamin in the market.

3 Growth performance, antioxidative pressure and immunological competence

High lean meat production is associated with chronic inflammation and activation of the innate immune system vis-à-vis cellular stress. This may negatively affect adaptive immune responses. It has been furthermore shown that lean muscle pigs show abnormally high serum concentrations of reactive oxygen metabolites, as opposed to rural swine (Brambilla et al., 2002). Amadori and Zanotti (2016) have experienced that reactive oxygen metabolite levels above 20 mM H_2O_2 can easily be found in lean pigs. This level implies an oxidative stress under resting conditions similar to human beings during intense physical stress. The high release of reactive oxygen metabolites in the presence of hypoxia is related to the mitochondrial electron transport chain. Moreover, there is strong evidence that the hypoxia response is highly detrimental to T-cell homeostasis, whereby Th2 and Th17 responses are favoured at the expense of Th1 and Treg ones (Amadori and Zanotti, 2016). Thus, lean-type pigs with high growth performance under conditions of chronic oxidative stress are not likely to induce balanced and effective T-cell responses because of a defective homeostasis of the T-cell compartment.

Oxidative stress accompanies infectious diseases and plays a dual role as free radicals to protect against invading microorganism, and they can also cause tissue damage during resulting inflammation. The mechanisms by which vitamins have antioxidant activity are addressed below. However, vitamins have direct and indirect influence on the pigs' immunocompetence beyond their antioxidative activity. Nutritional immunology seeks to increase or modulate the immune response through manipulation of the level of dietary nutrients. Many reports have described how vitamins can modulate cytokine production after *in vitro* or *in vivo* supplementation (Hernandez et al., 2009); however, little research on this topic has been performed on pigs.

3.1 Fat-soluble vitamins

The fat-soluble vitamins A, D and E seem to be important for immunomodulation: Vitamin A controls immunity through retinoid acid signalling, and of special interest is the retinoic acid synthesis in the gut (for review, see Guo et al., 2015). Many studies in human and other animal species have demonstrated the interaction between vitamin A and immune competence in relation to infectious diseases. Vitamin A enhanced Th2 cytokines and improved the immune response against gastrointestinal parasites (Dawson et al., 2006; Wang et al., 2007). Cells of the immune system express the vitamin D receptor and activated macrophages to produce 1,25 dihydroxyvitamin D_3 (1,25(OH)$_2$$D_3$). Without sun exposure or dietary vitamin D supplementation, pigs may have a

$25(OH)_2D_3$ level that is too low to support a high level of $1,25(OH)_2D_3$ synthesis by activated macrophages, which may compromise the immunoregulatory functions at the sites of inflammation. Recent research related to humans has supported the role of the active form of vitamin D ($1,25(OH)_2D_3$) in promoting normal function of innate and adaptive immune systems (Szymczak and Pawliczak, 2016), and this calls for further studies in pigs.

Vitamin E also has immunomodulatory effects, and some effects are attributed to the reversal of deleterious influence of reactive oxygen intermediates on immune functions like other antioxidants (e.g. ascorbic acid and selenium), whereas supplementation may augment antibody response. Results of Hernandez et al. (2009) revealed that vitamin E is able to suppress IL-10 production and to influence the production of IL-2, IL-4 and maybe TBX21 after *in vitro* stimulation in peripheral blood mononuclear cells isolated from healthy pigs. Table 1 gives an overview of studies performed on swine with the aim of studying vitamin E supplementation and effect on immunity of swine. Most of the studies addressed the influence on humoral immune responses. As can be deduced from Table 1, some studies reported an effect of vitamin E supplementation on humoral and/or cell-mediated immunity, while others did not observe any effect. However, the responses on the immune system in relation to vitamin supplementation may depend on the vitamin status and body reserves, and the exposure to stress (e.g. weaning and pathogen challenge) or other conditions under which the experiment is conducted. For example, we observed a dramatic reduction of the hepatic α-tocopherol content after infection challenge of pigs with *Escherichia coli* (Lauridsen et al., 2011), and using pigs that are hereditary deficient for ascorbate synthesis has demonstrated the influence of vitamin C on pigs' immune competence. The biological activity of vitamin E in relation to immunity of pigs should probably be seen and understood in light of protection of polyunsaturated fatty acids and membrane qualities that polyunsaturated fatty acids bring about. In our study (Møller and Lauridsen, 2006), dietary fatty acid composition influenced immune responses of macrophages after *E. coli* stimulation *ex vivo*, but dietary vitamin E supplementation, which influenced cellular α-tocopherol concentration, had no influence on the cytokine production of alveolar macrophages.

3.2 Water-soluble vitamins

Folates, vitamin B$_{12}$ and the intermediary amino acid homocysteine. Homocysteine is a sulphur-containing amino acid derived from the hydrolysis of S-adenosylhomocysteine (SAH) generated from S-adenosylmethionine (SAM), the major cellular methyl donor (Fig. 2). Among more than hundreds of enzymatic methylation reactions mediated by SAM, there are DNA methylation which control gene transcription and genetic stability, and synthesis of

Table 1 Effect of vitamin E supplementation and immunity

Animals	Vitamin E supplementation	Effect of immunity	Reference
Weaners	11, 110, 220 IU/kg	No effect on humoral immune responses	Blodgett et al., 1988
Weaners	11 vs. 220 IU/kg	Increase in antibody titre	Peplowski et al., 1981
Weaners	11, 110, 220 IU/kg	No effect on humoral immune responses	Blodgett et al., 1988
Weaners	11 vs. 220 IU/kg	No effect on antibody response to SRBC (or performance)	Bonnette et al., 1990a
Weaners	11, 110, 220, 550 IU/kg feed (at two housing temperatures/19 vs. 30°C)	No effect on humoral and cell-mediated immune response	Bonnette et al., 1990b
Gilts	13, 48 and 136 mg/kg. After weaning, pigs fed with the same diets	Immunized weaned pigs suckling sows on high vitamin E had higher AO to ovalbumin	Babinszky et al., 1991
Gilts/sows	22, 44, 88 IU/kg (gestation) and 55, 110, 220 IU/kg (lactation)	No effect on sows' immunity, but newborn pigs from high E levels (110 and 220 IU/kg) had higher lymphocyte response to PHA and Con A than pigs from sows fed 55 IU/kg	Nemec et al., 1994
Piglets	Injection (500 IU) i.m.	Higher AO to KLH in piglets injected with vitamin E from d 21 after birth	Hidiroglou et al., 1995
Sows/ piglets	70, 150 or 250 mg/kg	AO response to *E. coli* in piglets affected when sows supplemented with vitamin E	Lauridsen and Jensen, 2005
Weaners	85, 150 and 300 IU/kg at varying fat composition	No effect of vitamin E on cell-mediated immune responses, but effect of fatty acid composition	Møller and Lauridsen, 2006

phosphatidylcholine, choline, creatine as well as several neurotransmitters. However, intense methylation reactions as expected in lean pigs frequently resulted in accumulation of homocysteine (Hoffman, 2011). Elevated levels of blood plasma homocysteine are correlated with several pathologies and are recognized as an initiating factor for arteriosclerosis of coronary, cerebral and peripheral vessels (Boushey et al., 1995; Refsum et al., 1998). They also have harmful effects on embryo development (Pietrzik and Bronstrup, 1997; DiSimone et al., 2004) and cell proliferation (Chen et al., 2000). In view of the adverse effects of homocysteine on tissue integrity and its pro-oxidizing

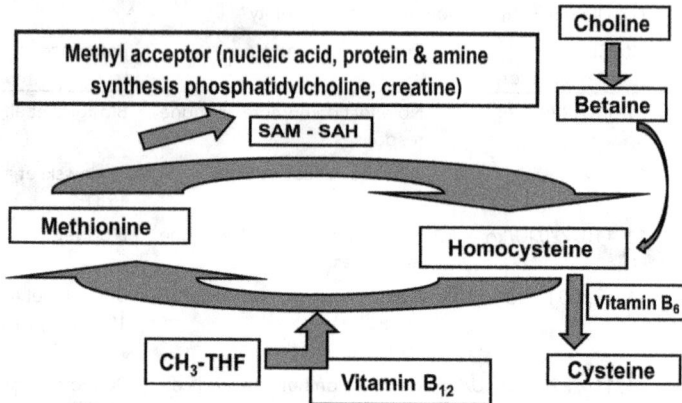

Figure 2 Transmethylation, remethylation and transsulphuration pathways of methionine. SAM, S-adenosylmethionine; SAH, S-adenosylhomocysteine; CH3-THF, methyltetrahydrofolate.

properties, organisms must rid themselves of this metabolite as quickly and as efficiently as possible. Homocysteine is metabolized via either remethylation (back to methionine) or transsulphuration (synthesis of cysteine) reactions. In remethylation, the conversion of homocysteine to methionine is catalysed by the enzyme methionine synthase, a vitamin B_{12}-dependent zinc protein (Bässler, 1997). This reaction requires the release of a methyl group from CH_3-H_4folate to H_4folate, the two main circulating forms of folic acid (vitamin B_9) in pigs (Natsuhori et al., 1996). Remethylation can also proceed through another zinc-containing enzyme, betaine-homocysteine methyltransferase. Betaine is an intermediate step in the catabolism of choline. In this last case, transmethylation capacity is limited because the tissue distribution of the enzyme is restricted to specific organs (Delgado-Reyes et al., 2001). In transsulphuration, the disposal of homocysteine is catalysed by the B_6-dependent cystathionine β-synthase, which leads to cystathionine and subsequently to cysteine and GSH biosyntheses. Therefore, removal of homocysteine is frequently impaired by a lack of folic acid, vitamin B_{12} or vitamin B_6 (Brosnan et al., 2007).

In growing and finishing pigs, even after vitamin supplementations, the plasma homocysteine concentration remained two to three times higher than the values observed in other species (<10 μM) such as dairy cows (Girard et al., 2005), rats (Sarwar et al. 2000), mice (Hofman et al., 2001), cats (Ruaux et al., 2001) and humans (Pietrzik and Brönstrup 1997). Some aspects of the remethylation pathway of methionine deserve to be further explored in pigs.

In piglets, at birth, blood plasma levels of homocysteine are very low (approximately 3 μM) and increase very rapidly at approximately 10 μM within 24 hours and at 18 μM and 30 μM at 7 and 14 days of age, respectively (Balance and House, 2005; Simard et al., 2007). During the last decade, we have showed in our laboratory that administrations of dietary supplements of folic acid and vitamin B_{12} to sows during gestation (Barkow et al., 2001; Guay et al., 2002; Simard et al., 2007) and lactation (Audet et al., 2015) can be used as efficient tools to increase maternal transfer (*in utero* or through colostrum and milk) of these vitamins to piglets which, in turn, prevents the rapid rise of homocysteine in blood plasma.

According to what was presented in the Introduction section about the evolution of hyperprolificity and the individual share of maternal vitamin transfer among piglet littermates during the perinatal period, it is likely that, during the last decades, it has included vitamins related to the chronic hyperhomocysteinaemia in piglets. Besides its multiple deleterious effects in several aspects of metabolism, homocysteine is also recognized as a powerful pro-oxidant which interferes with the immune system in stimulating, for example, the ubiquitous transcription factor NF-κB, a key regulator of pro-inflammatory cytokines such as TNF-alpha and IFN-gamma (Huang et al., 2001; Wang et al., 2000). Nevertheless, the question remained as to whether this atypical early regulation of homocysteine metabolism with high-plasma homocysteine is harmful to piglets or inversely, whether low-plasma homocysteine is beneficial. In this way, two populations of piglets (birth to 8 weeks of age) with high or low homocysteinaemia were generated by altering the transfer of folates and B_{12} from dams and by direct intramuscular (i.m.) injections of B_{12} to piglets using typical supplies of these vitamins (Audet et al., 2015). Growth performance, antibody and cellular immune responses, and some indicators of homocysteine metabolism, were then assessed according to vitamin treatments and their effects on plasma homocysteine. Indicators of cell-mediated immunity suggested a weakened immune competence with high homocysteinaemia (Audet et al., 2015). No treatment effect was reported on growth performance, but correlations were observed between plasma homocysteine and growth rate and feed conversion (feed:gain). However, these correlations were unexpectedly positive, a response which changed completely the perspective of the link between homocysteine and the immune system. Therefore, the only logical interpretation to such results would be that young 'high-performing' piglets which also generate high levels of plasma homocysteine are immunologically more fragile (Audet et al., 2015) and logically less resistant to disease challenges. The concept is important and may become a plausible explanation to the opposite relation often reported in concrete husbandry situations between superior performance and disease resistance. This side effect of high performance would be also coherent with the apparent global fragility of piglets against antigenic pressure which has developed during the last 30 years in several outbreaks of major diseases.

In order to verify if additional reductions of growth-related homocysteinaemia are possible in suckling piglets, a further study was undertaken with other nutrients related to metabolic disposal of homocysteine (Côté-Robitaille et al., 2015). The attempt was made to regulate plasma homocysteine in suckling piglets by daily oral supplementations of betaine recognized as a methyl group supplier, creatine for reducing the global demand for methyl groups coming from methionine, choline for combining both previous functions of betaine and creatine, and vitamin B_6 as an enzymatic co-factor for homocysteine catabolism. Concentrations of plasma homocysteine were reduced by 23% with the combination of all nutrients, such as betaine, pyridoxine, choline and creatine, supplied daily directly to suckling piglets during lactation. Nevertheless, the amplitude of this response on homocysteinaemia was similar to the strategy of supplementation with folic acid and vitamin B_{12} used by Audet et al. (2015). Additional studies are needed to know if a combination of approaches used by Côté-Robitaille et al. (2015) and Audet et al. (2015) could act in synergy to further enhance the clearance rate of homocysteine in suckling and post-weaned piglets and improve eventually their robustness in terms of health status.

4 Vitamins and antioxidation capacity: new perspectives

It is well known that reactive oxygen species (ROS) toxicity is controlled by complex networks of non-enzymatic (notably vitamins E and C) and enzymatic antioxidants, including the selenium (Se)-dependent glutathione peroxidase (GPX) system as well as copper (Cu), zinc (Zn)-dependent super-oxide dismutase system. In order to protect cells from oxidative reactions, vitamin E (α-tocopherol) needs to be incorporated in between the fatty acid methyl esters of the phospholipids in the cellular membranes. α-Tocopherol is mostly concentrated in membrane-rich fractions such as mitochondria and microsomes, and as reviewed by Lauridsen and Jensen (2012), it is possible by dietary means to further increase the concentration. Thus, incorporation of α-tocopherol and other antioxidants into mitochondria and other cellular compartments is important in order to maintain oxidative stability of the membrane-bound lipids and prevent damage from the ROS, and this is of major importance both for the living animal for disease prevention and for storage and processing of muscle meat.

Due to the interest in finding alternatives to the use of animal fat in the production of meat, studies on growing pigs have been conducted in order to understand the interaction between dietary fatty acid quality and antioxidants on markers of oxidative stress and antioxidant status as well as growth and organ function. More recently, the elevated cost and greater use of feed grains for biofuel production have increased the use of by-products with greater concentrations of polyunsaturated fatty acids, enhancing the potential for nutrient oxidation and oxidative stress in pig. Diets highly concentrated with PUFA as found in vegetable oils such as soya bean and linseed oils have increased potential for nutrient oxidation. Synthetic antioxidants such as butylated hydroxytoluene (BHT) and ethoxyquin are commonly used to preserve lipids and vitamins in feed and to protect the animal from oxidative stress. Lu et al. (2014a,b) reported that barrows fed with diets with an antioxidant blend (ethoxyquin and propyl gallate) or this blend with 11 IU/kg feed of vitamin E-attenuated oxidative stress in pigs provided a high-oxidant diet. However, the protective effects of supplemental vitamin E at 11 IU/kg relative to the antioxidant blend alone were minimal. In this way, Lauridsen et al. (1999) showed that a difference in 11 IU/kg (basic diet vs a basic diet supplemented with 6% rapeseed oil) actually influenced the oxidative status of muscles in pigs, as muscles of pigs on the rapeseed oil-based diets were less prone to oxidation, probably because of the content of vitamin E. However, increased susceptibility to oxidative stress because of processing (heating and mincing) of the pork enhances the requirement for antioxidant protection (Jensen et al., 1998). Thus, as also reported recently by Preveraud et al. (2014), supplementation with PUFA markedly decreases the concentration of α-tocopherol in the blood and tissues of growing pigs, and the ratio of α-tocopherol:PUFA needs to be carefully considered to meet the vitamin E requirements for pigs to ensure an optimal α-tocopherol meat enrichment. Recently, we showed (Lauridsen et al., 2013) that supplementation of PUFA (5% fish oil or sunflower oil) reduced the concentration of RRR tocopherol in plasma and muscles compared to saturated fatty acids (animal fat). In practice, it is recommended to supplement, per kg of diet, with 5 mg of vitamin E for each 10 g of dietary fat over a basic level of 30 g (Isabel et al., 2013) and that for all categories of pigs (recommended vitamin level: 100–150 IU/kg of vitamin E for piglets, sows and boars). However, this may vary depending on the proportion of PUFA in the dietary fat considering that the oxidative susceptibility increases with a factor of 2400 (from C18:0 to C18:3).

Recent focus on pig research has also been given to the interaction of antioxidative vitamins with other micronutrients, and especially with Se. In a recently published review (Surai and Fisinin, 2016), the role of selenium and its interaction with vitamins E and C as part of the antioxidant system of the body is described and explained in relation to the effect of dietary selenium (organic source vs inorganic) supplementation for protection against oxidative stress in sows and newly born piglets. The importance of storage of selenium in pig muscles and tissues to be used in stress conditions is addressed herein.

The glutathione peroxidase system (oxidized and reduced glutathione and the two enzymes, glutathione peroxidase and reductase) as mentioned above is closely related to the metabolism of the two amino acids, cysteine and methionine (Fig. 2), via the key junction metabolite, homocysteine. The *trans*sulphuration reaction provides a direct link between homocysteine and glutathione, the major redox buffer in mammalian cells. In this pathway, as mentioned previously, the disposal of homocysteine is catalysed by the B_6-dependent cystathionine β-synthase which leads to cystathionine and subsequently to cysteine and glutathione biosyntheses (Fig. 2). In fact, the two major homocysteine-utilizing enzymes, methionine synthase and cystathionine β-synthase, show reciprocal sensitivity to the oxidative conditions (Mosharov et al., 2000). In fact, reactive oxygen metabolites and peroxides upregulate the B_6-related transsulphuration pathway (homocysteine towards cysteine and eventually the glutathione peroxidase system) and downregulate the folate-B_{12}-related transmethylation (homocysteine towards methionine) (Fig. 2). Therefore, the presence (or the production) of oxidative metabolites will stimulate the pathway (transsulphuration) which will eventually neutralize them. The balance between remethylation and transsulphuration pathways during an oxidative stress regulates not only the flow of sulphur methionine and sulphur cysteine for provision of glutathione but also the flow of their seleno-methionine and seleno-cysteine for the GPX system. In fact, the animal metabolism does not distinguish between methionine and cysteine (sulphur forms) and their selenium analogues (cited by Daniels, 1996). In this case, seleno-cysteine also becomes a key factor along with seleno-homocysteine. Seleno-cysteine can be transformed in seleno-glutathione, but the size of that metabolic pool as compared to sulphur-glutathione is likely to be negligible (Fig. 2, small arrows). The importance of seleno-cysteine comes from its role for gene expression and activity of the Se-dependent enzyme, glutathione peroxidase (Flohé et al., 1997; Johansson et al., 2005). This role is due to a metabolite of seleno-cysteine called seleno-cysteinyl (a molecular form of cysteine moiety), which allows the full gene transcription of the enzyme (Johansson et al., 2005). In this process, one could think that the seleno-cysteine is cleaved and directly incorporated in the process, but this is not so (Fig. 2) as the metabolism involved is much more complicated: in fact, seleno-cysteine is mineralized through a metabolic step with the enzyme seleno-cysteine lyase (SCLY), a B_6-dependent enzyme specific to Se-cysteine which controls the flow of organic selenium towards the GPX system. The selenide produced from this reaction is then transformed into selenium phosphate before being resynthesized in the seleno-cysteinyl which is metabolically active for Se-dependent glutathione peroxidase (Yasumoto et al., 1979; Suzuki and Ogra, 2002; Johansson et al., 2005) (Fig. 2). It appears, therefore, that oxidative pressure will drive the B_6-dependent transsulphuration and 'transselenization' to produce, respectively, more GSH and further activate GPX.

Most of the reactions involved in the fate of seleno-methionine, seleno-homocysteine and seleno-cysteine for the control of the glutathione peroxidase system (Fig. 2) are B_6-dependent (Yasumoto et al., 1979). A previous report on rats has demonstrated that the response of Se-GPX to organic selenium could be dependent upon the dietary provision of vitamin B_6 (Yasumoto et al., 1979) (Fig. 3). In normal redox balance with the addition of dietary organic selenium, Se-GPX was lower at low B_6 status than after dietary supplement of vitamin B_6 (Fig. 4). The status of vitamin B_6 could play a critical role for an adequate flow of organic selenium towards the glutathione peroxidase system and might have to be taken into account for a reliable interpretation of the different studies with selenium supplementations. This last concept was addressed recently (Bueno-Dalto et al., 2015) using gilts during the peri-oestral period as a model promoting consistent and reliable oxidative stress conditions. Indeed, it is recognized that metabolic by-products such as reactive oxygen metabolites are generated during the peri-oestrus period in response to drastic enhancement of ovarian cellular energy metabolism in relation to hormone production and ovulation (Agarwal et al., 2005). Such production of reactive oxygen metabolites is particularly important for polytocous species such as pigs and probably more for modern hyperprolific lines of sows with high ovulation rates. In these animals, the oxidative pressure at the ovary level is, in fact, important enough to be detectable on enzymatic antioxidants like GPX activity at peripheric level (whole blood) during the peri-oestral period (Fig. 5) (Fortier et al., 2012). In fact, during that peri-oestral period, gene expressions of *GPX1* and *SCLY* were substantially higher in liver from gilts supplemented with organic Se and B_6 as compared to organic Se without B_6 (Bueno-Dalto et al., 2015) (Fig. 6). These last two effects confirm the crucial role of vitamin B_6 in modulating the metabolic pathway of organic Se towards the GPX system and the importance of adequate levels of this vitamin for an appropriate response of GPX during episodes of oxidative pressure. This last metabolic condition appears particularly critical because B_6 and Se responses on gene expressions of these key enzymes could not be confirmed when measurements were made at a different physiological stage without any particular surges of oxidative pressure (Bueno-Dalto et al., 2016).

5 Conclusion and future trends

As mentioned in the Introduction section, the criteria used to estimate vitamin requirements for pigs have changed, during the past century, from prevention of deficiencies to optimization of performance for growth and reproduction. Such criteria can be quantitatively assessed rather easily. As presented in this chapter, the estimation of requirements for the future will have to include criteria related to other aspects potentially limiting for full expression of performances such as metabolic stress and disease resistance. This means that vitamin requirements might be 'adjustable' depending upon the risk of fragility of the animal vis-à-vis critical periods or disseminated contaminants.

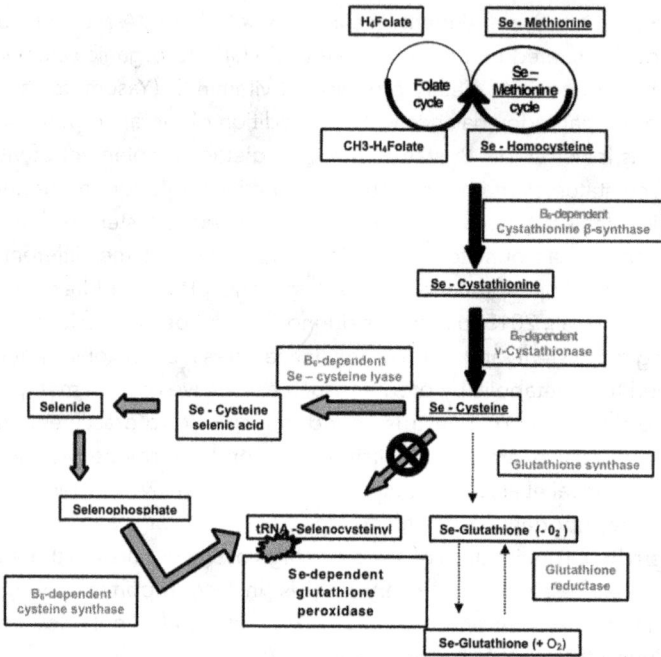

Figure 3 Se-amino acids metabolism and fate of Se–cysteine (mineralization) towards activation of Se-dependent glutathione peroxidase system.

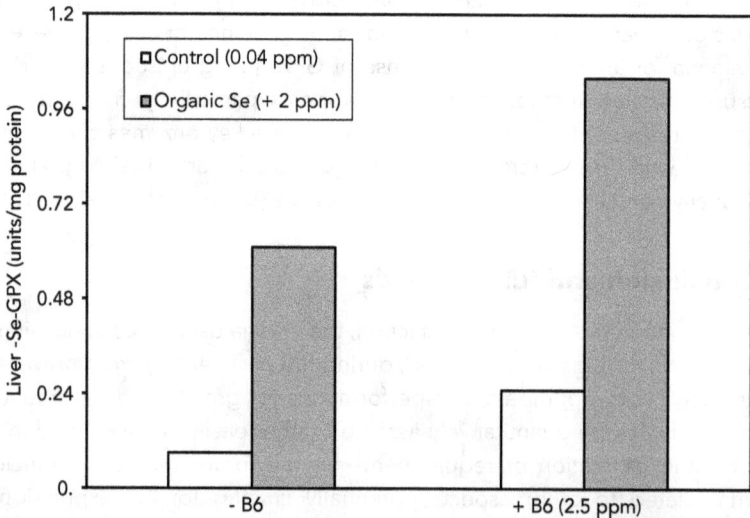

Figure 4 Effect of vitamin B_6 status on the hepatic response of Se-glutathione peroxidase (Se-GPX) to the dietary form of selenium. Adapted from Yasumoto et al. (1979).

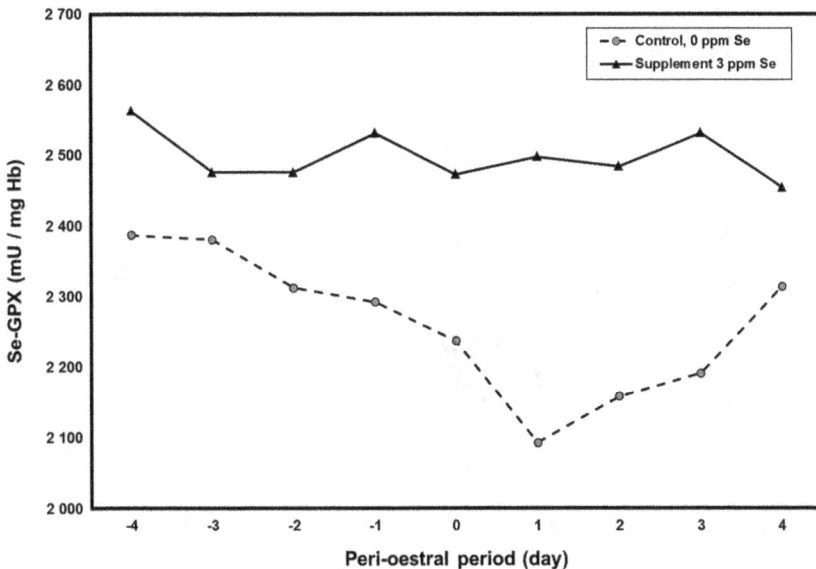

Figure 5 Blood Se-dependent glutathione peroxidase (Se-GPX) activity during the peri-oestrus period in gilts according to the level of dietary Se. The control diet contained a basal level of 0.2 mg/kg Se (unsupplemented), whereas the supplemented one represented a pooled profile from the control diet complemented with an extra 0.3 mg/kg (of inorganic selenite (MSe) or organic Se (yeast)). Day 0 = third oestrus. One unit (U) of GPX activity equals 1 mmol of NADPH oxidized per minute. Adapted from Fortier et al. (2012).

Besides the two levels of requirements for prevention of vitamin deficiencies and optimization of performance, there would be another one, likely at a higher level for prophylactic robustness of animals. Such a new approach to look at vitamin allowances in pigs nevertheless presents a future challenge because reliable indicators to assess the levels of metabolic stress and disease protection are still lacking.

6 Where to look for further information

The authors have previously provided an overview of vitamins and vitamin utilization in swine (see Matte and Lauridsen (2013) in the references for further details). The book chapter integrated the recent scientific advances in vitamin nutrition of pigs in an attempt to complement the basic information available in various publications on nutrient recommendation of swine (NRC, BSAS, ARC).

The authors have furthermore contributed with proceeding papers on specific vitamins in connection with seminars held by American Society of Animal Science (ASAS), Animal Nutrition Association of Canada, European

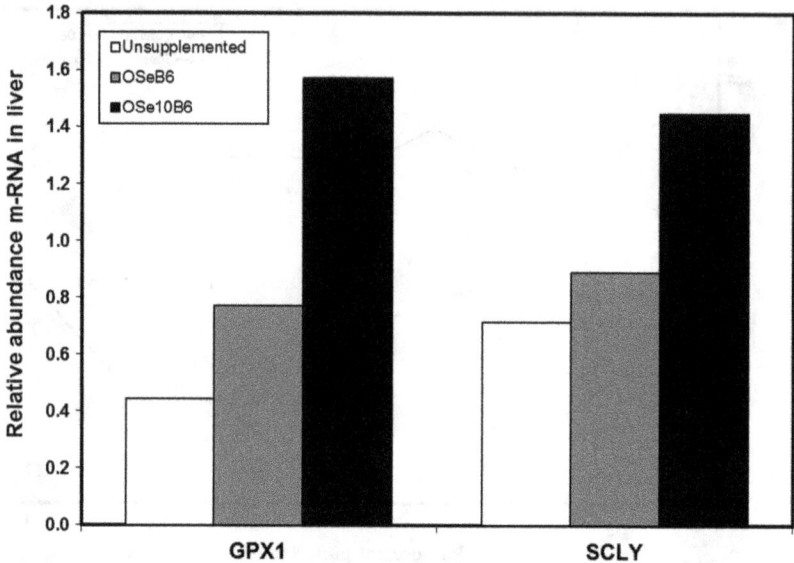

Figure 6 Gene expressions of Se-dependent glutathione peroxidase 1 (*GPX1*) and selenocysteine lyase (*SCLY*) in liver of gilts 3 days after the fourth post-pubertal oestrus. Unsupplemented = basal diet containing 0.2 mg/kg and 1.7 mg/kg of native Se and pyridoxine respectively, OSeB60 = basal diet supplemented with 0.3 mg/kg of Se-enriched yeast, OSeB610 = basal diet supplemented with 0.3 mg/kg of Se-enriched yeast and 10 mg/kg of hydro-chloride pyridoxine. Adapted from Bueno-Dalto et al. (2015).

Federation of Animal Science and European Society of Veterinary and Comparative Nutrition, e.g. Lauridsen (2013, 2014) and Matte (2014, 2015).

- Vitamins and vitamin utilization in swine/Matte, J. J. and Lauridsen, C. 2013. *Sustainable Swine Nutrition*. red. / Lee I Chiba. 1. udg. Wiley-Blackwell, s. 139-72.
- Update on Vitamin C, D, E and fatty acids in swine nutrition/Lauridsen, C. 2013. Proceedings of 49th Eastern Nutrition Conference. Animal Nutrition Association of Canada, Quebec, Cananda, s. 143-54.
- Triennial Growth Symposium - Establishment of the 2012 vitamin D requirements in swine with focus on dietary forms and levels of vitamin D/ Lauridsen, C. 2014. Journal of Animal Science, 92 (3), s. 910-16.
- Water-soluble vitamins and reproduction in sows: beyond prolificacy/ Matte, J. J. 2014. Book of Abstracts of the 65th Annual Meeting of the European Federation of Animal Science. Wageningen Academic Publishers, The Netherlands, No. 20, p. 108.

- New concepts on micronutrient requirements for sows: beyond reproductive performance/Matte, J. J. Proceeding of ESVCN 2015, Toulouse, France.
- Micronutrient requirements for sows: to be revisited for an adequate transfer to piglets? /Matte, J. J. 2015. Proceedings of Western Nutrition Conference, Animal Nutrition Association of Canada, Winnipeg, Manitoba, Pre-Conference Session.

7 References

Abedelahi, A., Salehnia, M., Allameh, A. A. and Davoodi, D. (2010), 'Sodium selenite improves the in vitro follicular development by reducing the reactive oxygen species leveland increasing the total antioxidant capacity and glutathione peroxide activity', *Hum. Reprod.*, 25, 977–85.

Agarwal, A., Gupta, S. and Sharma, R. K. (2005), 'Role of oxidative in stress in female reproduction', *Reprod. Biol. Endocrinol.*, 3, 1–28.

Al-Gubory, K. H., Fowler, P. A. and Garrel, C. (2010), 'The roles of cellular reactive oxygen species, oxidative stress and antioxidants in pregnancy outcomes', *Int. J. Biochem. Cell*, 42, 1634–50.

Amadori, M. and Zanotti, C. (2016), 'Immunoprophylaxis in intensive farming systems: the way forward', *Vet. Immunol. Immunopath.*, doi: 10.1016/j.vetimm.2016.02.011. [Epub ahead of print].

Audet, I., Girard, C. L., Lessard, M., Lo Verso, L., Beaudoin, F. and Matte, J. J. (2015), 'Homocysteine metabolism, growth performance, and immune responses in suckling and weanling piglets', *J. Anim. Sci.*, 93, 147–57.

Ayuso, M., Óvilo, C., Fernández, A., Nuñnez, Y., Isabel, B., Daza, A., López-Bote, C. J. and Rey, A. I. (2015), 'Effects of dietary vitamin A supplementation or restriction and its timing on retinol and α-tocopherol accumulation and gene expression in heavy pigs', *Anim. Feed. Sci. Tech.*, 202, 62–74.

Babinszky, L., Langhout, D. J., Verstegen, M. W., den Hartog, L. A., Joling, P. and Nieuwland, M. (1991), 'Effect of vitamin E and fat source in sows' diets on immune response of suckling and weaned piglets', *J. Anim. Sci.*, 69, 1833–42.

Ballance, D. M. and House, J. D. (2005), 'Development of the enzymes of Hcy metabolism from birth through weaning in the pig', *J. Anim. Sci.*, 83 (Suppl. 1), 160.

Barkow, B., Pietrzik, K. and Flachowsky, G. (2001), 'Effect of folic acid supplements on homocysteine in plasma of gestating sows', *Arch. Anim. Nutr.*, 54, 81–85. https://doi.org/10.1080/17450390109381967

BASF (Badische Anilin- & Soda-Fabrik) Corp. (1993), 'Vitamin Supplementation Rates for U.S. Commercial Poultry, Swine and Dairy Cattle' BASF Corp., Mount Olive, NJ.

BASF (Badische Anilin- & Soda-Fabrik) Corp. (2001), 'Dietary fortifications in vitamins for pigs and poultry in Canada, Technical seminar' BASF Corp., Toronto, ON & Richelieu, QC, Canada.

Bässler, K. H. (1997), 'Enzymatic effects of folic acid and vitamin B_{12}', *Internat. J. Vit. Nutr. Res.*, 67, 385–8.

Blodgett, D. J., Kornegay, E. T. and Schurig, G. G. (1998), 'Vitamin E-Selenium and immune response to selected antigens in swine', *Nutr. Rep. Intern.*, 38, 37–43.

Bonnette, E. D., Kornegay, E. T., Lindemann, M. D. and Notter, D. R. (1990a), 'Influence of two supplemental vitamin E levels and weaning age on performance, humoral antibody production and serum cortisol levels of pigs', *J. Anim. Sci.*, 68, 346-1353.

Bonnette, E. D., Kornegay, E. T., Lindemann, M. D. and Hammerberg, C. (1990b), 'Humoral and cell-mediated immune response and performance of weaned pigs fed four supplemental vitamin E levels and housed at two nursery temperatures', *J. Anim. Sci.*, 68, 1337-45.

Boushey, C. J., Beresford, S. A., Omenn, G. S. and Motulsky, A. G. (1995), 'A quantitative assessment of plasma homocysteine as a risk factor for vascular disease. Probable benefits of increasing folic acid intakes', *JAMA*, 274, 1049-57.

Brambilla, G., Civitareale, C., Ballerini, A., Fiori, M., Amadori, M., Archetti, L. I., Regini, M. and Betti, M. (2002), 'Response to oxidative stress as welfare parameter in swine', *Redox. Rep.*, 7, 159-63.

Brosnan, J. T., Brosnan, M. E., Bertolo, R. F. P. and Brunton, J. A. (2007), 'Methionine: A metabolically unique amino acid', *Livestock Sci.*, 112, 2-7.

Bueno-Dalto, D., Roy, M., Audet, I., Palin, M.-F., Guay, F., Lapointe, J. and Matte, J. J. (2015), 'Interaction between vitamin B6 and source of selenium on the response of the selenium-dependent glutathione peroxidase system to oxidative stress induced by oestrus in pubertal pig', *J. Trace Elem. Med. Biol.*, 32, 21-9.

Bueno-Dalto, D., Audet, I., Lapointe, J. and Matte J. J. (2016), 'The importance of pyridoxine for the impact of the dietary selenium sources on redox balance, embryo development, and reproductive performance in gilts', *J. Trace Elem. Med. Biol.*, 34, 79-89.

Burrild, A., Lauridsen, C., Faqir, N., Sommer, H. M. and Jakobsen, J. (2016), 'Vitamin D3 and 25-hydroxyvitamin D3 in pork and their relationship to vitamin D status in pigs', *J. Nutr. Sci.*, 5, e3.

CCSI (Canadian Centre for Swine Improvement) (2015), Annual Report, 36 p. https://www.ccsi.ca/main.cfm?target_page=annual.

Chen, C., Halkos, M. F., Surowiec, S. M., Conklin, B. S., Lin P. H. and Lumsden, A. B. (2000), 'Effects of homocysteine on smooth muscle cell proliferation in both cell culture and artery perfusion culture models', *J. Surg. Res.*, 88, 26-33.

Chew, B. P. (1993), 'Effects of supplemental β-carotene and vitamin A on reproduction in swine', *J. Anim. Sci.*, 71, 247-52.

Côté-Robitaille, M.-E., Girard, C. L., Guay, F. and Matte, J. J. (2015), 'Oral supplementations of betaine, choline, creatine and vitamin B6 and their influence on the development of homocysteinaemia in neonatal piglets', *J. Nutr. Sci.* 4(e31), 1-7. https://doi.org/10.1017/jns.2015.19

Daniels, L. A. (1996), 'Selenium metabolism and bioavailability', *Biol. Trace Element Res.*, 54, 185-99.

Dawson, H. D., Collins, G., Pyle, R., Key, M., Weeraratna, A., Deep-Dixit, V., Nadal, C. N. and Taup, D. D. (2006), 'Direct and indirect effects of retinoic acid on human Th2 cytokine and chemokine expression by human T lymphocytes', *BMC Immunol.*, 7, 207.

Delgado-Reyes, C. V., Wallig, M. A. and Garrow, T. A. (2001), 'Immunohistochemical detection of betaine-homocysteine S-methyltransferase in human, pig, and rat liver and kidney', *Archiv. Biochem. Biophys.*, 393, 184-6.

DiSimone, N., Riccardi, P., Maggiano, N., Piacentani, A., D'Asta, M., Capelli, A. and Caruso, A. (2004), 'Effect of folic acid on homocysteine-induced trophoblast apoptosis', *Molecul. Hum. Reprod.*, 10, 665-9.

Farquharson, J. and Adams, J. F. (1976), 'The forms of vitamin B_{12} in foods', Br. J. Nutr., 36, 127-36.

Fie, M., Zee, J. A. and Amiot, J. (1994), 'Séparation et quantification des isomères de la vitamine B_{12} dans le lait et certains produits laitiers par chromatographie liquide haute performance et par radio-essai', Sciences des aliments 14, 763-75.

Flohr, J. R., DeRouchey, J. M., Woodworth, J. C., Tokach, M. D., Goodband, R. D. and Dritz, S. S. (2016), 'A survey of current feeding regimens for vitamins and trace minerals in the US swine industry', J. Swine Health Prod., 24, 290-303.

Flohé, L., Wingender, E. and Brigelus-Flohé, R. (1997), Regulation of glutathione peroxidase. In Oxidative Stress and Signal Transduction. Forman, H. J. and Cadenas, H. ed., Chapman and Hall, New York, USA, pp. 415-40.

Fortier, M. E., Audet, I., Giguère, A., Laforest, J. P., Bilodeau, J. F., Quesnel, H. and Matte, J. J. (2012), 'Effect of dietary organic and inorganic selenium on antioxidant status, embryo development and reproductive performance in hyperovulatory first-parity gilts', J. Anim. Sci., 90, 231-40.

Giguère, A., Girard, C. L. and Matte, J. J. (2008), 'Methionine, folic acid and vitamin B-12 in growing-finishing pigs: Impact on growth performance and meat quality', Arch. Anim. Nutr., 62, 193-206. https://doi.org/10.1080/17450390802027494

Guay, F., Matte, J. J., Girard, C. L. Palin, M. F. Giguère, A. and Laforest, J. P. (2002), 'Effects of folic acid and vitamin B_{12} supplements on folate and homocysteine metabolism in pigs during early pregnancy', Br. J. Nutr., 88, 253-63.

Guo, Y., Brown, C., Ortiz, C. and Noelle, R. J. (2015), 'Leukocyte homing, fate, and function are controlled by retinoic acid', Physiol. Rev., 95, 125-48.

Herbert, V. (1988), 'Vitamin B-12: plant sources, requirements, and assay,' Am. J. Clin. Nutr., 48, 852-8.

Hernandez, J., Soto-Canevett, E., Pinelli-Saavedra, A., Resendiz, M., Moya-Cararena, S. Y. and Klasing, K. C. (2009), 'In vitro effect of vitamin E on lectin-stimulated porcine peripheral blood mononuclear cells', Vet. Immunol. Immunopath., 131, 9-16.

Hidiroglou, M., Batra, T. R., Farnworth, E. R. and Markham., F. (1995), 'Effect of vitamin E supplementation on immune status and α-tocopherol in plasma of piglets', Reprod. Nutr. Dev., 35, 443-50.

Hofmann, M. A., Lalla, E., Lu, Y., Gleason, M. R., Wolf, B. M., Tanji, N., Ferran Jr., L. J., Khol, B., Rao, V., Kisiel, W., Stern, D. M. and Schmidt, A. M. (2001), 'Hyperhomocysteinemia enhances vascular inflammation and accelerated atherosclerosis in a murine model', J. Clin. Invest., 107, 675-83.

Huang, R. F. S., Hsu, Y. C., Lin, H. L. and Yang, F. L. (2001), 'Folate depletion and elevated plasma homocysteine promote oxidative stress in rat livers', J. Nutr., 131, 33-8.

Isabel, B., Rey, A. I. and Lopez Bote, C. (2013), Optimum Vitamin Nutrition - In the production of quality animal foods. 5M Publishing Benchmark House, Sheffield, United Kingdom.

Jensen, C., Lauridsen, C. and Bertelsen, G. (1998), 'Dietary vitamin E: Quality and storage stability of pork and poultry', Trends Food Sci. Technol., 9, 62-72.

Johansson, L. G., Gafvelin, G. and Arner, E. S. J. (2005), 'Selenocysteine in proteins - properties and biotechnological use', Biochim. Biophys. Acta Gen. Subj., 1726, 1-13.

Kovacs, C. S. (2008), 'Vitamin D in pregnancy and lactation: Maternal, fetal, and neonatal outcomes from human and animal studies', Am. J. Clin. Nutr., 88(Suppl.), 520S-8S.

Lauridsen, C. and Jensen, S. K. (2005), 'Influence of supplementation of all-rac-alpha-tocopheryl acetate preweaning and vitamin C postweaning on alpha-tocopherol and immune responses of piglets', *J. Anim. Sci.*, 83, 1274–86.

Lauridsen, C. and Jensen, S. K. (2007), 'Transfer of vitamin E in milk to the newborn.' In *The Encyclopedia of Vitamin E* (eds. V. R. Preedy and R. R. Watson). Chapter 47, CAB International.

Lauridsen, C. and Jensen, S. K. (2012), 'α-Tocopherol incorporation in mitochondria and microsomes upon supranutritional vitamin E supplementation', *Genes Nutr.*, doi: 10.1007/s12263-012-0286-6.

Lauridsen, C., Engel, H., Jensen, S. K., Craig, A. M. and Traber, M. G. (2002a), 'Lactating sows and suckling piglets preferentially incorporate RRR- over all-rac-α-tocopherol into milk, plasma, and tissues', *J. Nutr.*, 132, 1258–64.

Lauridsen, C., Engel, H., Craig, A. M. and Traber, M. G. (2002b), 'Relative bioactivity of dietary RRR- and all-rac- rac-α-tocopheryl acetates in swine assessed with deuterium-labeled vitamin E', *J. Anim. Sci.*, 80, 702–7.

Lauridsen, C., Larsen, T., Halekoh and Jensen, S. K. (2010), 'Reproductive performance and bone status markers of gilts and lactating sows supplemented with two different forms of vitamin D', *J. Anim. Sci.*, 88, 202–13.

Lauridsen, C., Nielsen, J. H., Henckel, P. and Sørensen, M. T. (1999), 'Antioxidative and oxidative status in muscles of pigs fed rapeseed oil, vitamin E and copper', *J. Anim. Sci.*, 77, 105–15.

Lauridsen, C., Theil, P. K. and Jensen, S. K. (2013), 'Composition of alpha-tocopherol and fatty acids in porcine tissues after dietary supplementation with vitamin E and different fat sources', *Anim. Feed Sci. Technol.*, 179, 93–102.

Lauridsen, C., Vestergaard, E. M., Højsgaard, S., Jensen, S. K. and Sørensen, M. T. (2011), 'Inoculation of weaned pigs with E. coli reduces depots of vitamin E', *Livest. Sci.*, 137, 161–7.

Le Grusse, J. and Watier, B. (1993), *Les vitamines. Données biochimiques, nutritionnelles et cliniques. CEIV.*, Neuilly-sur-Seine, France.

Lu, T., Harper, A. F., Zhao, J., Estienne, M. J. and Dalloul, R. A. (2014a), 'Supplementing antioxidants to pigs fed diets high in oxidants: I. Effects on growth performance, liver function, and oxidative status', *J. Anim. Sci.*, 92, 5455–63.

Lu, T., Harper, A. F., Dibner, J. J. Scheffler, J. M., Corl, B. A., Estienne, M. J. Zhao, J. and Dalloul, R. A. (2014b), 'Supplementing antioxidants to pigs fed diets high in oxidants: II. Effects on carcass characteristics, meat quality, and fatty acid profile', *J. Anim. Sci.*, 92, 5464–75.

Matte, J. J., Guay, F. and Girard, C. L. (2012), 'Bioavailability of vitamin B 12 in cows' milk', *Brit. J. Nutr.*, 107, 61–6.

Matte, J. J., Guay, F., Le Floc'h, N. and Girard, C. L. (2010), 'Bioavailability of dietary cyanocobalamin (vitamin B_{12}) in growing pigs', *J. Anim. Sci.* 88, 3936–44.

Matte, J. J., Audet, I., Ouattara, B., Bissonnette, N., Talbot, G., Lapointe, J., Guay, F., Lo Verso, L. and Lessard, M. (2016), 'Sources and routes of administration of copper and vitamins A and D on metabolic status of these micronutrients in suckling piglets', *J. Anim. Sci.*, 94 (Suppl. 2), 113.

Matte J. J. and Lauridsen, C. (2013), 'Vitamins and vitamin utilization in swine.' In Chiba, Lee L. (Eds), *Sustainable Swine Nutrition*, Book chapter 6, ISUP, John Witey & Sons, Inc. Publ., 2121 State, Ames, Iowa 50014 USA.

Matte, J. J., Audet, I. and Girard, C. L. (2014a), 'Le transfert périnatal des vitamines et minéraux mineurs de la truie à ses porcelets: au-delà d'une seule insuffisance en fer?' *Journées Rech. Porcine en France*, 46, 71-6.

Matte, J. J., Audet, I. and Girard, C. L. (2014b), 'The perinatal transfer of vitamins and trace elements from sows to piglets', *J. Anim. Sci.*, 92. (Suppl 2), 153.

Matte, J. J., Audet, I., Ouattara, B., Bissonnette, N., Talbot, G., Lapointe, J., Guay, F., Lo Verso, L. and Lessard, M. (2017), 'Sources et voies d'administration du cuivre et des vitamines A et D sur le statut métabolique de ces micronutriments chez les porcelets sous la mère.' *Journées Rech. Porcine en France*, 49, (in press).

McGlone, J. J. (2000), 'Deletion of supplemental minerals and vitamins during the late finishing period does not affect pig weight gain and feed intake', *J. Anim. Sci.*, 78, 2797-2800.

Møller, S. and Lauridsen, C. (2006), 'Dietary fatty acid composition rather than vitamin E supplementation ex vivo cytokine and eicosanoid response of porcine alveolar macrophages', *Cytokine*, 35, 1-2, 6-12. https://doi.org/10.1016/j.cyto.2006.07.001

Mosharov, E., Cranford, M. R. and Banerjee, R. (2000), 'The quantitatively important relationship between homocysteine metabolism and glutathione synthesis by the transsulfuration pathway and its regulation by redox changes', *Biochemistry*, 39, 13005-11.

Natsuhori, M., Shimoda, M. and Kokue, E. (1996), 'Alteration of plasma folates in gestating sows and newborn piglets', *Am. J. Physiol.*, 270, R99-104.

Nemec, M., Butler, G., Hidiroglou, M., Farnworth, E. R. and Nielsen, K. (1994), 'Effect of supplementing gilts' diets with different levels of vitamin E and different fats on the humoral and cellular immunity of gilts and their progeny',. *J Anim. Sci.*, 72, 665-76.

Olivares, A., Daza, A., Rey, A. I. and Lopez-Bote, C. J. (2009), 'Interactions between genotype, dietary fat saturation and vitamin A concentration on intramuscular fat content and fatty acid composition in pigs', *Meat Sci.*, 82, 6-12.

Peplowski, M. A., Mahan, D. C, Murray, F. A., Moson, A. L., Canior, A. H. and Ekstrom, K. E. (1981), 'Effect of dietary and injectable vitamin E and selenium in weanling swine antigenically challenged with sheep red blood cells', *J. Anim. Sci.*, 51, 344.

Pietrzik, K. and Bronstrup, A. (1997), 'Folate in preventive medicine: A new role in cardiovascular disease, neural tube defects and cancer', *Ann. Nutr. Metab.*, 41, 331-43.

Prévéraud, D. P., Devillard, E., Rouffineau, F. and Borel, P. (2014), 'Effect of the type of dietary triacylglycerol fatty acids on α-tocopherol concentration in plasma and tissues of growing pigs', *J. Anim. Sci.*, 92, 4972-80. https://doi.org/10.2527/jas.2013-7099

Refsum, H., Ueland, P. M., Nygård, O. and Vollset, S. E. (1998), 'Homocysteine and cardiovascular disease', *Review. Annu. Rev. Med.*, 49, 31-62.

Ruaux, C. G., Steiner, J. M. and Williams, D. A. (2001), 'Metabolism of amino acids in cats with severe cobalamin deficiency', *Am. J. Vet. Res.*, 62, 1852-8.

Sahlin, A. and House, J. D. (2006), 'Enhancing the vitamin content of meat and eggs: Implications for the human diet', *Can. J. Anim. Sci.*, 86, 181-95.

Sarwar, G., Peace, R. W., Botting, H. G., L'Abbé, M. R. and Keagy P. M. (2000), 'Influence of dietary methionine with or without adequate dietary vitamins on hyperhomocysteinemia in rats', *Nutr. Res.*, 20, 1817-27.

Scott, J. M. (1997), 'Bioavailability of vitamin B12', *Eur. J. Clin. Nutr.*, 51, S49-53.

Seo, E.-G., Einhorn, T. A. and Norman, A. W. (1997), '24R,25-Dihydroxyvitamin D3: An Essential Vitamin D3 Metabolite for Both Normal Bone Integrity and Healing of Tibial Fracture in Chicks', *Endocrinol.*, 138, 3864-72. https://doi.org/10.1210/en.138.9.3864

Shaw, D. T., Rozeboom, D. W., Hill, G. M., Booren, A. M. and Link, J. E. (2002), 'Impact of vitamin and mineral supplement withdrawal and wheat middling inclusion on finishing pig growth performance, fecal mineral concentration, carcass characteristics, and the nutrient content and oxidative stability of pork', *J. Anim. Sci.*, 80, 2920-30.

Simard, F., Guay, F., Girard, C. L., Giguère, A., Laforest J. P. and Matte, J. J. (2007), 'Effects of concentrations of cyanocobalamin in the gestation diet on some criteria of vitamin B_{12} metabolism in the first-parity sows', *J. Anim. Sci.*, 85, 3294-302.

Surai, P. F. and Fisinin, V. I. (2016), 'Selenium in sow nutrition', *Anim. Feed Sci. Nutr.*, 211, 18-30.

Szymczak, I. and Pawliczak, R. (2016), 'The active Metabolite of Vitamin D3 as a Potential Immunomodulator', *Scand. J. Immunol.*, 83, 83-91. https://doi.org/10.1111/sji.12403

Theil, P. K., Lauridsen, C. and Quesnel, H. (2014), 'Neonatal piglet survival: Impact of sow nutrition around parturition on fetal glycogen deposition and production and composition of colostrum and transient milk', *Animal*, 8, 7, 1021-30.

VSP. (2016), Notat nr.1611. Danish Pig Research Centre. SEGES, Copenhagen, Denmark.

Wang, G., Siow, Y. L. and Karmin, O. (2000), 'Homocysteine stimulates nuclear factor (B activity and monocyte chemoattractant protein-1 expression in vacular smooth-muscle cells: A possible role for protein kinase C', *Biochem. J.*, 352, 817-26.

Wang, X., Allen, C. and Ballow, M. (2007), 'Retinoic acid enhances the production of IL-10 while reducing the synthesis of IL-12 and TNF-alpha from LPS-stimulated monocytes/macrophages', *J. Clin. Immunol.*, 27, 193-200.

Yasumoto, K., Iwami, K. and Yoshida, M. (1979), 'Vitamin B6 dependence of selenomethionine and selenite utilization for glutathione peroxidase in the rat', *J. Nutr.*, 109, 760-6.

Chapter 5

Developing seaweed/macroalgae as feed for pigs

Marta López-Alonso, Universidade de Santiago de Compostela, Spain; Marco García-Vaquero, University College Dublin, Ireland; and Marta Miranda, Universidade de Santiago de Compostela, Spain

1 Introduction

The global human population is projected to reach 9.7 billion by 2050. It has been estimated that overall food production will have to double by 2050 to meet the increased demand (Aiking, 2014). Animal farming systems will have to increase production by an estimated 70%, increasing the demand for animal feed by up to 235% (Herman and Schmidt, 2016). In meeting the increasing demand, intensive production systems have traditionally focused on maximising production outputs while minimising production costs (Jones et al., 2012). High-intensity farming has been characterised by genetically selecting animals for fast growth and high meat content, kept at high stocking densities and fed nutrient-dense plant-based feeds (Swanson, 1995). Potential consequences have been the greater levels of stress for animals and their vulnerability to disease, leading to the increased prophylactic use of antibiotics in some cases (Van Boeckel et al., 2015).

It has been increasingly recognised that some aspects of these production systems are no longer sustainable. Researchers have begun to investigate alternative feed sources that can be more sustainably produced than

http://dx.doi.org/10.19103/AS.2021.0091.15

conventional, land-based crops used for feed, and which also have the potential to boost animals' natural immunity to disease whilst meeting their nutritional requirements, thus reducing the reliance on antibiotics (García-Vaquero, 2018; Miranda et al., 2017; Morais et al., 2020).

Macroalgae or seaweeds comprise a diverse group with more than 10 000 different species described to date (Collins et al., 2016). Approximately 5% of them are currently exploited for feed applications (Michalak and Chojnacka, 2014). They are able to adapt to changing and extreme marine environmental conditions by producing secondary metabolites including lipids, polysaccharides and minerals, to protect the biomass from cell damage caused by these marine stressors (Garcia-Vaquero et al., 2017). These metabolites make them a promising potential feed ingredient.

The use of macroalgae as a source of protein and other essential nutrients (such as polyunsaturated fatty acids, phenolic compounds and minerals) in animal nutrition are promising dietary strategies to complement traditional feed sources (Øverland et al., 2019). The incorporation of macroalgal extracts containing carbohydrates such as laminarin and fucoidan has, for example, shown promising results as an alternative to antibiotics in piglets, either directly by feed supplementation and/or indirectly via maternal dietary supplementation (Sweeney and O'Doherty, 2016). This chapter summarises the nutritional properties of macroalgae for use in animal feed, both as a source of protein and as antimicrobial compounds in swine production systems.

2 Challenges in using macroalgae for feed applications

The use of macroalgae to feed livestock has been documented for thousands of years since at least the time of Ancient Greece (Makkar et al., 2016). In the nineteenth and early twentieth centuries, seaweed was used in the coastal regions of France, the Scottish islands and Scandinavia, mainly to feed ruminants but also other livestock species, including pigs (Chapman, 2012; Sauvageau, 1920). The use of brown algae like *Ascophyllum nodosum* or *Fucus vesiculosus* in pig nutrition has been documented since the beginning of the twentieth century. Macroalgae were traditionally mixed with cereal meals to fatten pigs in countries in Northern Europe (Chapman, 2012; Evans and Critchley, 2014). The interest in this feedstock increased during periods of food shortage such as World War I, but usage declined in the first half of the twentieth century due to the perceived poor nutritional quality of the biomass (Makkar et al., 2016). The low levels of protein and energy available for metabolism, and the high mineral content of brown seaweeds led to their replacement by other protein ingredients such as fishmeal and soybean meal in formulated feed for monogastric animals (Øverland et al., 2019). It was also found that high amounts of brown seaweed could be harmful to health: the inclusion of 10% of *A. Nodosum* in pig feed over several weeks, for example, has been related to weight loss (Jones et al., 1979). To maximise the potential benefits of the inclusion of seaweeds in feeds for improved pig production and health, seaweeds can only be used at low concentrations in feed (1–2%) rather than as a major source of macronutrients (Morais et al., 2020).

As previously mentioned, the high amount of ash in macroalgae has hindered the inclusion of intact or dried macroalgae in animal diets, as feeds high in minerals have resulted in diarrhoea and decreased animal performance in pigs and poultry (Koreleski et al., 2010; Wilkinson, 1992). There has been a recent focus on refining biomass to reduce unwanted mineral content in favour of the more beneficial compounds.

When extracting polysaccharides, their molecular weight, monosaccharide composition and sulphate contents will change depending on the extraction and purification methods used, affecting the biological properties and effects of these molecules (Garcia-Vaquero et al., 2017). Ale and Meyer (2013) attributed the lack of official approval of polysaccharides or their derived fractions for pharmaceutical, dermatological, nutraceutical or other commercial applications to the lack of standardised extraction methodologies.

Recent research has focused on identifying the optimum extraction parameters such as temperature, time and pH (Garcia-Vaquero et al., 2019; Yuan et al., 2018). Yuan et al. (2018) extracted sulphated polysaccharides from the green macroalga *Ulva prolifera* using microwave-assisted hydrothermal extraction. The sulphur content increased with the temperature and acid concentration of the solvent. Antioxidant and pancreatic lipase inhibition properties were also influenced by extraction conditions due to structural modifications of the active molecules (Yuan et al., 2018). Garcia-Vaquero et al. (2019) used hydrothermal-assisted extraction and found significant variation in the levels of fucoidan, laminarin and antioxidant activity under differing conditions. The authors determined the optimum process parameters for a combined maximised extraction of these compounds from *Laminaria hyperborea* (120°C, 80.9 min and 12.02 volume of solvent-to-seaweed ratio). Other techniques, such as the combination of ultrasound and microwave technologies, have also been explored for brown macroalga *A. nodosum*, achieving increased yields of fucoidan (Garcia-Vaquero et al., 2020).

Techniques such as enzyme-assisted extraction have also attracted attention due to their high efficiency and mild extraction conditions that ensure less modification in the chemical structure of the compounds of interest (Michalak and Chojnacka, 2014). Enzymes have been used to extract multiple compounds including proteins, phenols, carotenoids and lipids (Billakanti et al., 2013; Wang et al., 2010). However, the use of these methods has been hindered by the high price of enzymes and the cost of scaling up production (Michalak and Chojnacka, 2014).

Improvements in the extraction of beneficial compounds have been one factor in renewed scientific interest in macroalgae as a source of multiple high-value compounds including proteins, polysaccharides, lipids, pigments and minerals with a wide variety of antioxidant, anti-bacterial, anti-tumour and other properties (Holdt and Kraan, 2011). This has made them promising ingredients for a wide variety of applications, for example, as nutraceuticals or functional foods, pharmaceuticals, cosmetics and in animal feed.

3 Composition of macroalgae

Based on their pigment composition, macroalgae can be classified as:

- brown macroalgae (Phaeophyta);
- red macroalgae (Rhodophyta); and
- green macroalgae (Chlorophyta).

Macroalgae are rich in carbohydrates, with medium or high amounts of proteins, low levels of lipids and variable mineral content (Dominguez and Loret, 2019; Kraan, 2013). The composition of macroalgae varies greatly depending on the macroalgal class, species, parts of the macroalgae sampled as well as the season and location (Øverland et al., 2019). The variable composition of macroalgae (water, ash, crude protein, crude lipids and polysaccharides) compiled from the recent scientific literature are summarised in Table 1.

As seen in Table 1, the moisture content of the macroalgal biomass is high, accounting in some cases for up to 94% of the weight of the collected fresh biomass. This means that once the biomass is collected, the common process is drying (freeze-drying or oven drying) to preserve the biomass, before applying further processing or technological treatments to obtain macroalgal

Table 1 Summary of the composition ranges described for Phaeophyta (brown macroalgae), Chlorophyta (green macroalgae) and Rhodophyta (red macroalgae) in the recent scientific literature. The content of this table was modified from the work of Øverland et al. (2019)

Proximate composition	Phaeophyta	Chlorophyta	Rhodophyta
Water (% water in wet biomass)	61-94	78-92	72-91
Ash (% DW basis)	15-45	11-55	12-42.2
Polysaccharides (% DW basis)	38-61	15-65	36-66
Crude protein (% DW basis)	2.4-16.8	3.2-35.2	6.4-37.6
Crude lipids (% DW basis)	0.3-9.6	0.3-2.8	0.2-12.9

DW, dry weight.

ingredients (Garcia-Vaquero et al., 2020; Rioux et al., 2010; Yuan et al., 2018). The following sections review in more detail the composition of macroalgae: protein, carbohydrates, lipids, minerals and other compounds. This is partly because variations in these compounds are a key challenge in the effective use of macroalgae in feed, and partly because their accurate characterisation is essential to the use of bioactive compounds with beneficial effects in pig nutrition.

3.1 Protein content in macroalgae

Protein content varies significantly, depending on multiple factors including species. Comparing protein accumulation between studies is difficult due to the different approaches used to estimate protein content in macroalgae. The most commonly used method is based on the quantification of the overall nitrogen content, estimating the value of crude protein by using multiple conversion factors. Due to the high amounts of non-protein nitrogen in macroalgae, several conversion factors have been calculated (Biancarosa et al., 2017; Makkar et al., 2016). The conversion factor used in each study affects the comparisons between studies.

In general, protein concentrations in brown macroalgae are lower (1-16%) compared to green (11-26%) and red (11-33%) macroalgae that have protein levels comparable to other traditionally used protein-rich products (Garcia-Vaquero and Hayes, 2016). The protein concentration is also influenced by season. The average protein content of Laminaria digitata, for example, is 6.8%, with the highest levels accumulated during the first quarter of the year (Schiener et al., 2015). Interspecies and intraspecies variations in protein content are shown in Table 2 (García-Vaquero, 2018). This table summarises the variable protein contents (1.1-26.7%) of selected brown, red and green macroalgae.

As shown in Table 2, the reported protein content of brown macroalgae varied between 1.1% (Nielsen et al., 2016) and 16.1% (Mols-Mortensen et al., 2017). Huge variations in the protein content have also been reported for red macroalgal species, with levels varying from 9.2 % (Chan and Matanjun, 2017) to 25.2% (Lozano et al., 2016). The protein levels described in green macroalgae varied between 11.2 % (Biancarosa et al., 2017) and 26.7% (Silva et al., 2015). A significant variation in the levels of protein has also been described within these macroalgal groups depending on the species and climatological conditions affecting the biomass. Studying compounds in European kelp varieties (Saccharina latissima and L digitata) collected in Denmark, Nielsen et al. (2016) reported the highest biomass production and protein content in seaweed from highly saline water, while low salinity resulted in higher amounts of fermentable sugars and pigments. Manns et al. (2017) reported that the variation in composition was mainly related to season, species and location. The

Table 2 Protein content in selected brown, red and green macroalgae species, originally published by García-Vaquero (2018) and reproduced with permission from Springer Nature

Macroalgae (sp.)	Protein content (% dry weight)	References
Phaeophyta (brown macroalgae)		
Fucus sp.	3.9–4.0	(Biancarosa et al., 2017)
Ascophyllum nodosum	3.0	(Biancarosa et al., 2017)
	6.8	(Schiener et al., 2015)
Laminaria sp.	1.5–5.1	(Nielsen et al., 2016)
	6.6	(Biancarosa et al., 2017)
	2.4–12.7	(Manns et al., 2017)
Saccharina latissima	1.1–7.5	(Nielsen et al., 2016)
	4.8–14.1	(Manns et al., 2017)
	4.0–16.1	(Mols-Mortensen et al., 2017)
	10.6	(Stévant et al., 2017)
Eisenia arborea	9.2–12.5	(Landa-Cansigno et al., 2017)
Alaria esculenta	10.6	(Stévant et al., 2017)
Rhodophyta (red macroalgae)		
Porphyra sp.	13.5–20.6	(Biancarosa et al., 2017)
Gracilaria sp.	9.2	(Chan and Matanjun, 2017)
	12.6	(Chan and Matanjun, 2017)
Chondrus crispus	11.0	(Biancarosa et al., 2017)
Palmaria palmata	10.6	(Biancarosa et al., 2017)
	10.9	(Schiener et al., 2017)
Pyropia columbina	25.2	(Lozano et al., 2016)
Chlorophyta (green macroalgae)		
Ulva sp.	26.7	(Silva et al., 2015)
	13.6	(Angell et al., 2017)
	11.2–15	(Biancarosa et al., 2017)
	22.5	(Bikker et al., 2016)
Cladophora rupestris	12.0	(Biancarosa et al., 2017)

environmental variables studied by the authors included temperature, salinity, phosphate, nitrate and ammonia. The authors reported high levels of protein (15–20% w/w) in seaweed samples collected during February and March from multiple locations, while the lowest amount of protein (2.3–8.4% w/w) was reported in the biomass collected during July and August (Manns et al., 2017). Research on *Saccharina* spp. from the Faroe Islands by Mols-Mortensen et al. (2017) found significantly higher protein content in biomass collected in April–May compared to July–August, while no significant differences were found between locations (Mols-Mortensen et al., 2017).

In contrast to carbohydrates, protein levels are highest during the winter and low during the summer (Adams et al., 2011; Fleurence, 1999; Schiener et al., 2015). It has been suggested that high levels of protein support the build-up of nitrogen reserves for rapid growth during the summer months (Chapman and Craigie, 1977). Schiener et al. (2015) reported the average contents of protein at 6.9 ± 1.1%, 6.8 ± 1.3%, 7.1 ± 1.7% and 11.0 ± 1.4% in *L. digitata*, *L. hyperborea*, *S. latissima* and *A. esculenta*, respectively. The contents of protein varied significantly through the year, with high levels of protein during the first quarter of the year for all macroalgal species, while the lowest protein contents were found in the third quarter of the year.

Amino acids combine to make proteins. Glutamic and aspartic acids are the most abundant amino acids in most macroalgae, with generally low levels of methionine (Holdt and Kraan, 2011). Brown macroalgae are considered rich sources of threonine, valine, leucine, lysine, glycine and alanine (Holdt and Kraan, 2011). Macroalgae usually contain high levels of glutamic acid, present both in its bound and free forms, contributing to the characteristic umami taste of algae (Mæhre et al., 2014). Lourenço et al. (2002) studied the amino acid and protein composition of 19 tropical seaweed species. The authors reported that green algae had low percentages of aspartic and glutamic acids, while red algae showed high percentages of lysine and arginine, and brown macroalgae had higher amounts of methionine. The authors also reported significant variations in the concentration of individual amino acids depending on macroalgal species. The highest concentration of glutamic acid (17.6% of total amino acids) was found in the brown macroalga *Sargassum vulgare*, while the lowest amount (10.7%) was reported in the green macroalga *Caulerpa fastigiate*. The amino acid content varied between species, location and season of collection. Manns et al. (2017) reported that the brown macroalgae *L. digitata* and *S. latissima* contained mainly glutamic acid, aspartic acid and alanine. Individual amino acid contents vary significantly between spring and summer, with levels more dependent on location than species. Glutamic acid represented over 25% of the total amino acid content of *L. digitata* collected in March, while these levels decreased to 12.9% in biomass collected in August. The highest levels of aspartic acid were found in August (14.5%) compared to March (9.5%).

The high levels of protein in macroalgae make it possible to produce bioactive peptides or cryptides. Bioactive peptides are sequences of 2–30 amino acids with no activity within the parent protein. After their release by several enzymatic hydrolysis or fermentation processes, these peptides have been linked to a wide range of health benefits in food and feed. Several bioactive peptides have been identified from *Undaria pinnatifida* (Suetsuna and Nakano, 2000). Some have been commercialised, such as the Wakame peptide jelly (Riken Vitamin Co., Ltd., Tokyo, Japan) and Nori peptide S (Shirako Co., Ltd., Tokyo, Japan) (Fukami, 2010).

3.2 Carbohydrates in macroalgae

Carbohydrates are one of the major components of macroalgae, with contents ranging from 4% to 76% depending on species, location and season (Holdt and Kraan, 2011). Macroalgal carbohydrates include a wide range of compounds (Lafarga et al., 2020). The major polysaccharides include alginate, carrageenan, and other phycocolloids such as agar, commonly used as stabilisers, thickeners and emulsifiers in food production. Red macroalgae mainly contain floridean starch, cellulose and other minor compounds such as mannans, xylans and sulphated galactans. Other carbohydrates, such as glucans (laminarin) and fucoidan, are mainly present in brown macroalgae, while green macroalgae are the main source of energy-reserve carbohydrates such as ulvans (Lafarga et al., 2020; Stiger-Pouvreau et al., 2016). As noted, the levels depend on factors such as species and season. Fucoidan levels in *F. vesiculosus*, for example, were higher than in *U. pinnatifida* (Holdt and Kraan, 2011). The highest amount of fucoidan in cultured *L. japonica* occurred when the biomass matured, rather than during the early stage of development (Honya et al., 1999).

Macroalgal carbohydrates such as fucoidan and laminarin have been associated with improved piglet performance and gut health (Gahan et al., 2009; McDonnell et al., 2010), and have been considered as alternatives to antibiotics and zinc oxide (Sweeney and O'Doherty, 2016). These polysaccharides demonstrate a range of anti-inflammatory, antioxidant, anti-coagulant, anti-viral and anti-tumour properties *in vitro* and *in vivo* (Garcia-Vaquero et al., 2017), with the potential to benefit human health through improving the antioxidant content of animal-derived products (Moroney et al., 2012, 2015). Dietary supplementation of extracts from brown seaweed *(L. digitata)* containing laminarin and fucoidan also improved the quality and shelf life of pig meat by lowering lipid oxidation (Moroney et al., 2012, 2015). Since polysaccharides such as alginates, fucoidan, and laminarin are not digested in the upper digestive tract of the animals, they can be considered as a source of dietary fibre for animals (Garcia-Vaquero et al., 2017). Lahaye (1991) estimated total dietary fibre ranging from 32.7% to 74.6% (on a DW basis) and water-soluble fractions ranging from 51.6% to 85%.

3.3 Lipids and other compounds in macroalgae

In general, macroalgae have a low lipid content, normally ranging between 0.2% and 12.9% as seen in Table 1. However, macroalgal lipids are rich in polyunsaturated fatty acids (PUFA), such as omega 6 and 3 (n6 and n3), with health benefits for both humans and animals. The relative abundance of a particular PUFA depends on the species. Red macroalgae normally have high concentrations of eicosapentaenoic acid (C20:5, n3) that can reach levels of up to half of the total fatty acids of the macroalgae; while brown macroalgae have in general high concentrations of oleic acid (C18:1, n9), linoleic acid (C18:2, n6) and α-linolenic acid (C18:3, n3) (Dawczynski et al., 2007).

Macroalgae are also rich in other phytochemicals including phenolic compounds, pigments and vitamins. As previously mentioned, the main macroalgal classes are established on the basis of the pigmentation of the biomass. Brown macroalgae are rich in fucoxanthin, while the colour of red macroalgae is associated with an abundance of phycobilins and other pigments such as chlorophylls (a and b) and carotenes. Xanthophylls are the main compounds influencing the pigmentation of green macroalgae (Øverland et al., 2019). Macroalgae also contain a wide range of phenolic compounds (i.e. phlorotannins and phloroglucinols) with potent antioxidant properties (Rajauria, 2018). Roleda et al. (2019) reported differences in phenolic compound levels depending on the species and season of collection. The levels of polyphenols in *Palmaria palmata* were significantly higher in Spring, compared to Summer and Autumn (Roleda et al., 2019).

3.4 Mineral contents in macroalgae

The amount of ash in the macroalgal biomass, ranging from 11% to 55% (Table 1), hinders the utilisation of full or intact dried macroalgae as animal feed. The incorporation of high levels of minerals in the feed of monogastric species, including pigs and poultry, resulted in diarrhoea and decreased animal performance (Koreleski et al., 2010; Wilkinson, 1992). There has been a recent study on the extraction of high-value compounds, concentrating them while decreasing the amount of minerals present in the raw biomass. Several innovative technologies, including ultrasound and microwaves, have been explored to obtain macroalgal ingredients or extracts rich in biologically active compounds, such as polysaccharides (i.e. fucoidan and laminarin), for feed applications (Garcia-Vaquero et al., 2017; Yuan et al., 2018).

Macroalgae can accumulate variable levels of minerals including essential trace elements (Ca, Co, Cr, Cu, Fe, I, Mg, Mn, Mo, Ni, P, Se and Zn), providing a source of essential minerals in animal feed (Rey-Crespo et al., 2014). The mineral levels vary by species due to varying metal-binding capacities (Güven et al., 1995). The cell wall polysaccharides produced by brown macroalgae have a high ability to absorb and retain metals. As a result, alginates produced by brown macroalgae have the strongest metal-binding capacity, followed by carrageenans and agar, predominantly produced by red macroalgae (Güven et al., 1995). Brown macroalgae have been described as being exceptionally high in iodine, compared to red and green macroalgae (Biancarosa et al., 2018; Mæhre et al., 2014).

Biancarosa et al. (2018) analysed the mineral profile of 21 macroalgal species collected from the coasts of Norway, and reported generally high iodine contents, varying from 22 mg/kg to 10 000 mg/kg dried weight (DW), with the highest accumulation reported in the brown macroalgae *L digitata*. Seaweed supplements in pigs are considered a way to increase the iodine content of meat (Dierick et al., 2009; He et al., 2002). Dierick et al. (2009) demonstrated that feeding pigs with 2% of *A. nodosum* increased the concentration of iodine in the tissue by 2.7 to 6.8, depending on the tissue.

Macroalgae are also known to accumulate toxic metals such as As, Cd and Hg. Yamada et al. (2007) reported different concentrations of Cd and Pb in *U pinnatifida* related to industrial waste. A number of regulators are addressing content of toxic metals in macroalgae when used as animal feed. In the European Union, calcareous marine algae must contain less than 10 mg As/kg and 15 mg Pb/kg relative to a feed, with moisture content of 12% (Commission Regulation 574/2011) (Regulation, 2011). A feedstuff containing macroalgae must contain less than 40 mg As/kg, with levels of the most toxic As species (inorganic As) of less than 2 mg/kg (Commission Regulation 574/2011) (Regulation, 2011). European legislation specifically mentions levels of As in the brown macroalga *Hizikia fusiforme*, as a result of previous reports of high levels (Besada et al., 2009; Rose et al., 2007).

Most studies analysing the mineral content of macroalgae have focused on analysing total As concentrations (Biancarosa et al., 2018; Khan et al., 2015; Ronan et al., 2017), even though As may be present in its organic form with little or no toxicity (Contam, 2009). Rose et al. (2007) reported that the level of inorganic As was less than 7% of total As. A recent study determining the mineral profile of three brown macroalgae (*L digitata, L hyperborea* and *A. nodosum*) collected in Ireland reported low levels of toxic metals (Cd, Hg and Pb), while levels of total As were high (49–64 mg/kg DW macroalgae) compared with previous reports (Garcia-Vaquero et al., 2021).

4 Biological functions and health-promoting effects of macroalgae and macroalgal-derived extracts in pig nutrition

Based on the properties of macroalgae, particularly brown seaweeds, a large body of recent research has been conducted in pigs. Table 3 summarises the experimental work conducted in the last decade on the use of macroalgae or macroalgal extracts in pigs.

There is limited information on the effects of dietary supplementation with full macroalgal biomass. Choi et al. (2017) described the beneficial effects on growth performance, gut microflora and intestinal morphology in weaning pigs receiving intact *Ecklonia cava* that could be attributed to the high levels of fucoidan (ca. 112 ± 6 g/Kg) in this brown macroalgae. However, Michiels et al. (2012) reported that the addition of *A. nodosum* had no effect on the

Table 3 Effects of macroalgae supplementation on the performance and gut health in weaned piglets. Supplemented diets were tested against a control diet

Compounds/Substances tested (doses)	Duration (days)	Growth performance	Nutrient digestibility	Gut microbiota	Gut architecture	Antioxidant activity	Immune system modulation	References
Macroalgal extracts								
Laminarin (300 mg/kg) and/or fucoidan (240 mg/kg) extracts	Weaning to 8 d			+	+		+	(Walsh et al., 2013a)
Brown seaweed (alginates oligosaccharides (50-200 mg/kg)	Weaning to 14 d	+	+			+	+	(Wan et al., 2017)
Brown seaweed (alginic acid oligosaccharides (100 mg/kg)	Weaning to 14 d	+			+	+	+	(Wan et al., 2018)
Fucoidan extract (44%), (125 and 250 mg/kg)	Weaning to 14 d	-	+					(Rattigan et al., 2019)
Laminarin extract (65%) (100, 200 and 300 mg/kg)	Weaning to 14 d	+		+				(Rattigan et al., 2020)
Laminarin extract (65%) (300 mg/kg)	Weaning to 14 d	+		+	+			(Vigors et al., 2020)
Brown seaweed (alginic acid oligosaccharides (100 mg/kg)	Weaning to 21 d	+	+	+	+	+	+	(Wan et al., 2016)
Laminarin (300 mg/kg) and/or fucoidan (240 mg/kg) extracts	Weaning to 21 d	+		+				(McDonnell et al., 2010)

(Continued)

Table 3 (*Continued*)

Compounds/Substances tested (doses)	Duration (days)	End-points*						References
		Growth performance	Nutrient digestibility	Gut microbiota	Gut architecture	Antioxidant activity	Immune system modulation	
Laminarin (300 mg/kg) and/or fucoidan (236 mg/kg) extracts	Weaning to 21 d	+		+				(O'Doherty et al., 2010)
Laminarin (300 mg/kg) and/or fucoidan (240 mg/kg) extracts	Weaning to 21 d	+	+					(McAlpine et al., 2012)
Seaweed extract containing laminarin (314 mg/kg) and fucoidan (249 mg/kg)	Weaning to 25 d	+	+	+				(Dillon et al., 2010)
Laminarin (300 mg/kg) and/or fucoidan (240 mg/kg) extracts	Weaning to 32 d	+	+	+				(Heim et al., 2014)
Laminarin (150–300 mg/kg) and/or fucoidan (240 mg/kg) extracts	Weaning to 35 d	+	+	+				(Walsh et al., 2013b)
Fucoidan extract (44%), (250 mg/kg) and laminarin extract (65%) (300 mg/kg)	Weaning to 35 d	-	+	+	+			(Vigors et al., 2021)
Laminarin (300 mg/kg) and/or fucoidan (240 mg/kg) extracts	Weaning to 40 d	+	+	+				(O'Shea et al., 2014)

Treatment	Duration						Reference
Commercial seaweed extract (OceanFeed Swine®) (5 g/kg)	Weaning to 160 d	+		+			(Ruiz et al., 2018)
Seaweed extract containing laminarin (sows 1 g/d; piglets 300 mg/kg) and fucoidan (sows 0.8 g/d; piglets 240 mg/kg)	107 d gestation-weaning (28 d)	+					(Draper et al., 2016)
Seaweed extract (10 g/d) containing laminarin (1 g) and fucoidan (0.8 g)	107 d gestation-weaning (28 d)				+	+	(Heim et al., 2015a)
Seaweed extract containing laminarin (280 mg/kg) and fucoidan (244 mg/kg)	107 d gestation-weaning (26 d)	+	−	+	−	+	(Leonard et al., 2011)
Seaweed extract (10 g/d) containing laminarin (1 g) and/or fucoidan (0.8 g)	107 d gestation-weaning (24 d)	+			+	+	(Heim et al., 2015b)
Whole macroalgae							
Ecklonia cava (0.5–1.5 g/kg)	Weaning to 28 d	+	−	+	+		(Choi et al., 2017)
Ascophyllum nodosum (2.5–10 g/kg)	Weaning to 28 d	−					(Michiels et al., 2012)

*Key of outputs: + (significant positive); − (significant negative) effect of the treatment compared to the control group; empty cell, not indicated or not determined.

performance or gut health and plasma oxidative status of piglets, possibly due to levels of bioactive compounds being too low to exert any prebiotic effect.

As shown in Table 3, many studies involve the use of macroalgal extracts in feed. The extracts studied mainly include mixtures of carbohydrates, such as laminarin and/or fucoidan, given as feed supplements to piglets during the post-weaning period (from weaning up to 40 days), since the weaning phase is a critical period due to the high incidence of enteric pathologies. Supplements have also been given to sows at the end of the gestation period, and the effects of this maternal supplementation on piglets were then analysed (Heim et al., 2015a,b; Leonard et al., 2011). Other studies have focused on the long-term effects of dietary supplementation fed to animals from weaning to slaughter (Draper et al., 2016; Ruiz et al., 2018).

4.1 Growth performance and digestion

The studies shown in Table 3 reported the overall positive effects on productive performance with the inclusion of macroalgal extracts in the diet. The average daily gain (ADG) of piglets fed brown seaweeds and/or macroalgal-derived ingredients was in all cases higher (ca. 5% to 40%) compared to that of piglets fed a control diet. Studies evaluating animals at slaughter found the effects of supplementation on ADG to be limited, with improvements of < 5 % compared to those receiving a control diet (Draper et al., 2016; Heim et al., 2015b). The positive effects of macroalgal extracts on growth seem to be related to the prebiotic effects of seaweed polysaccharides on piglet gut function and health (Corino et al., 2019). Overall, the inclusion of macroalgal extracts had positive effects on the digestibility of nitrogen (up to 8% increase), gross energy (up to 10% increase), fibre (up to 73% increase) and ash (up to 80% increase) in most of the studies. The improvement in nutrient digestibility seems to be related to the effect of carbohydrates and antioxidants on gut microbiota and on the villous architecture, with an increase in absorptive capacity and nutrient transporters (Sweeney and O'Doherty, 2016), and also to the beneficial effect on the intestinal mucosal cells and volatile fatty acid production (i.e. butyric acid) (Corino et al., 2019).

4.2 Gut function

Macroalgal supplements also have significant effects on gut microbiota (Table 3). Algal supplements stimulate the growth of *Lactobacilli* and reduce the population of Enterobacteriaceae such as *Escherichia coli*. Dierick et al. (2010) observed that 1% of *A. nodosum* added to feed reduced *E. coli*, while increasing the Lactobacilli/*E. coli* ratio, leading to a lower susceptibility of the animals to intestinal disorders.

Several studies have also demonstrated the positive influence of extracts from brown seaweed on the morphological features of the small intestine of suckling piglets, with increased villus height in the ileum. These positive effects have been attributed to an upregulation in the expression of the tight junction proteins, occludin (OCLN) and zonula occludens 1 (ZO-1) in the small intestine (Wan et al., 2016), together with enhanced antioxidant capacity, decreased mast cell degranulation and prevention of the release of pro-inflammatory cytokines via restraining the TLR4/NF-kB and NOD1/NF-kB signalling pathways (Wan et al., 2016, 2017, 2018). Maternal dietary supplementation with macroalgal-derived polysaccharides also down-regulated the gene expression of pro-inflammatory cytokines involved in *E. coli*-caused diarrhoea in piglets at 48 h after birth and weaning (Heim et al., 2015a).

A number of recent studies have focused on the effects of feeding fucoidan and laminarin for 14 days post-weaning (Rattigan et al., 2019; Vigors et al., 2020). Rattigan et al. (2020) found that laminarin supplementation at 300 ppm provided the most beneficial effects, with improvements in animal performance and positive effects in small intestinal morphology, microbial populations and gene expression. When analysing the microbiome of the laminarin-supplemented group, the changes observed included an increased abundance of *Prevotella* and reductions in *Enterobacteriaceae* (Vigors et al., 2020). Laminarin supplementation had a positive influence on intestinal health through alterations in the gastrointestinal microbiome and increased butyrate production (Vigors et al., 2021). However, there was no additional benefit on performance with either laminarin or fucoidan supplementation up to day 35 post-weaning.

4.3 Immune function and antioxidant capacity

Seaweed extracts also seem to play an immunomodulatory role. Leonard et al. (2011) reported that the dietary supplementation of sows from 109 days of gestation until weaning with extracts from the brown macroalga *Laminaria spp.* increased levels of immunoglobulins (Ig) G and A in the colostrum of sows by 19% and 25%, respectively, as well as the levels of IgG in the serum of piglets by 10%. Feeding seaweed extracts during lactation suppressed pro-inflammatory IL-1a mRNA expression in the ileum of pigs 11 days after weaning. Dietary seaweed extract supplementation post-weaning also induced up-regulation in colonic MUC2 mRNA expression in pigs 11 days post-weaning. These results demonstrate that feed supplementation post-weaning can improve gut health and growth performance in starter pigs.

Other macroalgal polysaccharides such as alginic acid supplemented at 100 mg/kg feed over 21 days post-weaning also increased IgG and IgA concentrations in piglet serum by 20% and 53%, respectively (Wan et al., 2016).

Alginic acid supplementation also increased superoxide dismutase, catalase and total antioxidant activities, and decreased the concentration of malonic dialdehyde in the serum of supplemented animals. The positive effects of alginic acid on the growth performance, antioxidant capacity, immunity and intestinal development in weaned pigs suggest that alginic acid could serve as a useful bioactive feed additive.

Other studies have evaluated the effect of dietary supplementation with laminarin and fucoidan on antioxidant activity (Moroney et al., 2012, 2015; Rajauria et al., 2016). Dietary supplementation of laminarin and fucoidan (0.45-0.9 g/kg) in pigs 3 weeks pre-slaughter resulted in enhanced meat due to deposition of marine-derived bioactive antioxidant components in muscle, reduced saturated fatty acids and lowered lipid oxidation in muscle (Moroney et al., 2015). This antioxidant response was attributed to mechanisms such as the reduction of saturated fatty acids and decreased lipid oxidation in the muscle, though it is still unclear if the free radical scavenging abilities of the extract are responsible for the antioxidant activity observed in the muscle. The enhanced meat quality effected by seaweed polysaccharides may be mediated through the health-promoting effects of gut-associated immunity. The improved fatty acid profiles and enhanced lipid stability of pork meat had no impact on the tenderness, flavour or other sensory properties of the meat. This suggests that dietary supplementation of seaweed extracts containing laminarin and fucoidan in pigs could result in an enhanced meat product. Rajauria et al. (2016) found an improvement in meat quality from the addition of a seaweed extract containing laminarin, due to decreased lipid oxidation. Improved antioxidant capacity and colour stability, together with the reduced lipid peroxidation in the muscle longissimus dorsi, suggest that feed supplementation with macroalgal polysaccharides may protect animals from oxidative stress-induced diseases and improve meat quality.

5 Conclusion and future trends

So far, beneficial effects have been mainly found when adding macroalgae at levels of ≤ 10% of the total concentration in animal feed. More studies of the biochemical profile of seaweed macro and micronutrients, and particularly seaweed bioactive metabolites are needed to properly calculate optimum rates of inclusion of the macroalgae in feed, both to optimise the beneficial effects and to avoid the negative or toxic effects of other compounds (i.e. As, Cd and Hg toxic metals). Wild seaweed biomass varies enormously in nutritional value because of seasonal geographical variations and varying risks of bioaccumulation of heavy metals. Regular monitoring is essential to provide a reliable source of safe animal feed supplementation. A large body of research has been conducted in the last decade on the supplementation of pigs' feed with macroalgal extracts, with beneficial effects related mainly to the presence of polysaccharides (laminarin and/or fucoidan) extracted from brown macroalgae. These studies demonstrate the potential of macroalgal compounds as feed supplements and their application in pig nutrition and production. Despite these promising results, particularly when used as substitutes for in-feed antibiotics, further research is needed to elucidate the effects of individual compounds present in macroalgal extracts. The production of extracts following optimised and standardised protocols, and the characterisation of the molecules within these extracts, will allow the establishment of clear structure–function relationships. This will allow industrial exploitation of macroalgal extracts for improved intestinal physiology, morphology, microbiology and immune response of post-weaning piglets, leading to industrial products with proven health-enhancement benefits that can be applied routinely in the herds.

6 Where to look for further information

As seen in the current chapter, macroalgae can be an excellent source of multiple compounds that can be used in animal feed for multiple purposes. However, the incorporation of macroalgae in animal feed will ultimately depend on the composition of the original feedstock. An introduction to the wide variation of the macroalgal biomass with respect to the species, season and year of collection can be accessed in Garcia-Vaquero et al. (2021). 'Seasonal variation of the proximate composition, mineral content, fatty acid profiles and other phytochemical constituents of selected brown macroalgae' (see full details in the reference list). This publication may serve as a reference guide for the inclusion of macroalgae in animal feed and the selective collection of this feedstock depending on the intended use in animals' diet. Moreover, one of the main organisations in Europe promoting the knowledge on algal biomass is the European Algae Biomass Association (EABA) (https://www.eaba-ass ociation.org/en/aboutus). This group include key information related to algal biomass and main key stakeholders of this industry in Europe.

Macroalgae, despite being an excellent source of multiple beneficial compounds, may also contain high levels of minerals and other minor compounds (heavy metals) that may limit the utilisation of this biomass either for practical purposes or legal issues related to the presence of these contaminants in the biomass. An extensive legislative framework on the use of macroalgae as animal feed in Europe can be accessed from Garcia-Vaquero and Hayes (2016) 'Red and green macroalgae for fish and animal feed and human functional food development' (see in the reference list). Furthermore, the use of macroalgae in animal feed is regulated in Europe by the European Food Safety Authority (EFSA) within the 'feed additives' panel. The readers of the chapter are referred to the EFSA and, in particular, to the publications of this panel to get the latest information related to the use of any fed additive (https://www.efsa.europa.eu/en/topics/topic/feed-additives). Individual countries may also provide their own guidance or interpretation of these European guidelines. As an example, the Food Safety Authority of Ireland (FSAI) published the report 'Safety Considerations of Seaweed and Seaweed-derived Foods Available on the Irish Market' (ISBN: 978-1-910348-42-0) and established recommendations for the seaweed sector in Ireland in addition to the aforementioned European regulations.

As the field of animal nutrition specialises, further studies will be needed not only including macroalgae or macroalgal extracts in animal feed, but targeting specific macroalgal molecules in animal feed to establish a clear association between the health benefits appreciated in the animals and particular macroalgal compounds that could be industrially exploited. In order to achieve that, multidisciplinary projects, including the optimisation of the extraction of macroalgal compounds as well as their *in vitro* and *in vivo* evaluation, should be further explored. Currently, the European project BIOCARB-4-FOOD (17RDSUSFOOD2ERA-NET1, https://www.biocarb4food.eu/) focuses on targeting the extraction and utilisation of macroalgal carbohydrates for multiple industrial applications. Other promising projects following also a multidisciplinary approach include the 'Enhance microalgae' (https://www.enhancemicroalgae .eu/). This European project offers a comprehensive approach to the study of algae for the generation of high-value compounds, starting from cultivation to extraction of these compounds from algae using novel technologies, as well as the final potential applications of these compounds in multiple industries.

7 Acknowledgements

This chapter did not receive any specific grant from funding agencies in the public, commercial or not-for-profit sectors.

8 References

Adams, J. M. M., Toop, T. A., Donnison, I. S. and Gallagher, J. A. (2011). Seasonal variation in Laminaria digitata and its impact on biochemical conversion routes to biofuels. *Bioresource Technology* 102(21): 9976-9984. DOI: 10.1016/j.biortech.2011.08.032.

Aiking, H. (2014). Protein production: planet, profit, plus people? *The American Journal of Clinical Nutrition* 100(Suppl. 1): 483S-489S. DOI: 10.3945/ajcn.113.071209.

Ale, M. T. and Meyer, A. S. (2013). Fucoidans from brown seaweeds: an update on structures, extraction techniques and use of enzymes as tools for structural elucidation. *RSC Advances*. Royal Society of Chemistry 3(22): 8131-8141. DOI: 10.1039/C3RA23373A.

Angell, A. R., Paul, N. A. and de Nys, R. (2017). A comparison of protocols for isolating and concentrating protein from the green seaweed Ulva ohnoi. *Journal of Applied Phycology* 29(2): 1011-1026. DOI: 10.1007/s10811-016-0972-7.

Besada, V., Andrade, J. M., Schultze, F. and González, J. J. (2009). Heavy metals in edible seaweeds commercialised for human consumption. *Journal of Marine Systems* 75(1-2): 305-313. DOI: 10.1016/j.jmarsys.2008.10.010.

Biancarosa, I., Espe, M., Bruckner, C. G., Heesch, S., Liland, N., Waagbø, R., Torstensen, B. and Lock, E. J. (2017). Amino acid composition, protein content, and nitrogen-to-protein conversion factors of 21 seaweed species from Norwegian waters. *Journal of Applied Phycology* 29(2): 1001-1009. DOI: 10.1007/s10811-016-0984-3.

Biancarosa, I., Belghit, I., Bruckner, C. G., Liland, N. S., Waagbø, R., Amlund, H., Heesch, S. and Lock, E. J. (2018). Chemical characterization of 21 species of marine macroalgae common in Norwegian waters: benefits of and limitations to their potential use in food and feed. *Journal of the Science of Food and Agriculture*. John Wiley & Sons, Ltd 98(5): 2035-2042. DOI: 10.1002/jsfa.8798.

Bikker, P., van Krimpen, M. M., van Wikselaar, P., Houweling-Tan, B., Scaccia, N., van Hal, J. W., Huijgen, W. J. J., Cone, J. W. and López-Contreras, A. M. (2016). Biorefinery of the green seaweed Ulva Lactuca to produce animal feed, chemicals and biofuels. *Journal of Applied Phycology*. Springer Netherlands 28(6): 3511-3525. DOI: 10.1007/s10811-016-0842-3.

Billakanti, J. M., Catchpole, O. J., Fenton, T. A., Mitchell, K. A. and MacKenzie, A. D. (2013). Enzyme-assisted extraction of fucoxanthin and lipids containing polyunsaturated fatty acids from Undaria pinnatifida using dimethyl ether and ethanol. *Process Biochemistry* 48(12): 1999-2008. DOI: 10.1016/j.procbio.2013.09.015.

Chan, P. T. and Matanjun, P. (2017). Chemical composition and physicochemical properties of tropical red seaweed, Gracilaria changii. *Food Chemistry* 221: 302-310. DOI: 10.1016/j.foodchem.2016.10.066.

Chapman, A. R. O. and Craigie, J. S. (1977). Seasonal growth in Laminaria longicruris: relations with dissolved inorganic nutrients and internal reserves of nitrogen. *Marine Biology* 40(3): 197-205. DOI: 10.1007/BF00390875.

Chapman, V. (2012). *Seaweeds and Their Uses*. New York: Springer Science & Business Media.

Choi, Y., Hosseindoust, A., Goel, A., Choi, Y., Hosseindoust, A., Goel, A., Lee, S., Jha, P. K., Kwon, I. K., & Chae, B. J. (2017). Effects of Ecklonia cava as fucoidan-rich algae on growth performance, nutrient digestibility, intestinal morphology and caecal microflora in weanling pigs. *Asian-Australasian Journal of Animal Sciences*. Asian-Australasian Association of Animal Production Societies (AAAP) and Korean Society of Animal Science and Technology (KSAST) 30(1): 64-70. DOI: 10.5713/ajas.16.0102.

Collins, K. G., Fitzgerald, G. F., Stanton, C. and Ross, R. P. (2016). Looking beyond the terrestrial: the potential of seaweed derived bioactives to treat non-communicable diseases. *Marine Drugs*. MDPI AG 14(3). DOI: 10.3390/md14030060.

Contam EP on C in the FC (2009) Scientific opinion on arsenic in food. *EFSA Journal* Wiley-Blackwell Publishing Ltd 7(10). DOI: 10.2903/j.efsa.2009.1351.

Corino, C., Modina, S. C., Di Giancamillo, A., Chiapparini, S. and Rossi, R. (2019). Seaweeds in pig nutrition. *Animals: An Open Access Journal from MDPI*. MDPI AG 9(12): 1126. DOI: 10.3390/ani9121126.

Dawczynski, C., Schubert, R. and Jahreis, G. (2007). Amino acids, fatty acids, and dietary fibre in edible seaweed products. *Food Chemistry* 103(3): 891-899. DOI: 10.1016/j.foodchem.2006.09.041.

Dierick, N., Ovyn, A. and De Smet, S. (2009). Effect of feeding intact brown seaweed Ascophyllum nodosum on some digestive parameters and on iodine content in edible tissues in pigs. *Journal of the Science of Food and Agriculture* 89(4): 584–594. DOI: 10.1002/jsfa.3480.

Dierick, N., Ovyn, A. and De Smet, S. (2010). In vitro assessment of the effect of intact marine brown macro-algae Ascophyllum nodosum on the gut flora of piglets. *Livestock Science* 133(1–3): 154–156. DOI: 10.1016/j.livsci.2010.06.051.

Dillon, S., Sweeney, T., Figat, S., Callan, J. J. and O'Doherty, J. V. (2010). The effects of lactose inclusion and seaweed extract on performance, nutrient digestibility and microbial populations in newly weaned piglets. *Livestock Science* 134(1–3): 205–207. DOI: 10.1016/j.livsci.2010.06.142.

Dominguez, H. and Loret, E. P. (2019). Ulva Lactuca, A source of troubles and potential riches. *Marine Drugs*. MDPI AG 17(6): 357. DOI: 10.3390/md17060357.

Draper, J., Walsh, A. M., McDonnell, M. and O'Doherty, J. V. (2016). Maternally offered seaweed extracts improves the performance and health status of the postweaned pig1. *Journal of Animal Science* 94(suppl_3): 391–394. DOI: 10.2527/jas.2015-9776.

Evans, F. D. and Critchley, A. T. (2014). Seaweeds for animal production use. *Journal of Applied Phycology* 26(2): 891–899. DOI: 10.1007/s10811-013-0162-9.

Fleurence, J. (1999). Seaweed proteins. *Trends in Food Science and Technology* 10(1): 25–28. DOI: 10.1016/S0924-2244(99)00015-1.

Fukami, H. (2010). *Functional Foods and Biotechnology in Japan*. DOI: 10.1201/9781420087123.

Gahan, D. A., Lynch, M. B., Callan, J. J., O'Sullivan, J. T. and O'Doherty, J. V. (2009). Performance of weanling piglets offered low-, medium- or high-lactose diets supplemented with a seaweed extract from Laminaria spp. *Animal* 3(1): 24–31. DOI: 10.1017/S1751731108003017.

Garcia-Vaquero, M. and Hayes, M. (2016). Red and green macroalgae for fish and animal feed and human functional food development. *Food Reviews International*. Taylor & Francis 32(1): 15–45. DOI: 10.1080/87559129.2015.1041184.

Garcia-Vaquero, M., Rajauria, G., O'Doherty, J. V. and Sweeney, T. (2017). Polysaccharides from macroalgae: recent advances, innovative technologies and challenges in extraction and purification. *Food Research International* 99(3): 1011–1020. DOI: 10.1016/j.foodres.2016.11.016.

García-Vaquero, M. (2018). Seaweed proteins and applications in animal feed. In: *Novel Proteins for Food, Pharmaceuticals and Agriculture*. Wiley Online Books, pp. 139–161. DOI: 10.1002/9781119385332.ch7.

Garcia-Vaquero, M., O'Doherty, J. V., Tiwari, B. K., Sweeney, T. and Rajauria, G. (2019). Enhancing the extraction of polysaccharides and antioxidants from macroalgae using sequential hydrothermal-assisted extraction followed by ultrasound and thermal technologies. *Marine Drugs*. MDPI AG 17(8): 457. DOI: 10.3390/md17080457.

Garcia-Vaquero, M., Ummat, V., Tiwari, B. and Rajauria, G. (2020). Exploring ultrasound, microwave and ultrasound–microwave assisted extraction technologies to increase the extraction of bioactive compounds and antioxidants from brown macroalgae. *Marine Drugs*. MDPI AG 18(3): 172. DOI: 10.3390/md18030172.

Garcia-Vaquero, M., Rajauria, G., Miranda, M., Sweeney, T., Lopez-Alonso, M. and O'Doherty, J. (2021). Seasonal variation of the proximate composition, mineral content, fatty acid profiles and other phytochemical constituents of selected brown macroalgae. *Marine Drugs* 19(4) . DOI: 10.3390/md19040204.

Güven, K. C., Akyüz, K. and Yurdun, T. (1995). Selectivity of heavy metal binding by algal polysaccharides. *Toxicological and Environmental Chemistry*. Taylor & Francis 47(1-2): 65-70. DOI: 10.1080/02772249509358127.

He, M. L., Hollwich, W. and Rambeck, W. A. (2002). Supplementation of algae to the diet of pigs: a new possibility to improve the iodine content in the meat. *Journal of Animal Physiology and Animal Nutrition* 86(3-4): 97-104. DOI: 10.1046/j.1439-0396.2002.00363.x.

Heim, G., Walsh, A. M., Sweeney, T., Doyle, D. N., O'Shea, C. J., Ryan, M. T. and O'Doherty, J. V. (2014). Effect of seaweed-derived laminarin and fucoidan and zinc oxide on gut morphology, nutrient transporters, nutrient digestibility, growth performance and selected microbial populations in weaned pigs. *British Journal of Nutrition*. Cambridge University Press 111(9): 1577-1585. DOI: 10.1017/S0007114513004224.

Heim, G., O'Doherty, J. V., O'Shea, C. J., Doyle, D. N., Egan, A. M., Thornton, K. and Sweeney, T. (2015a). Maternal supplementation of seaweed-derived polysaccharides improves intestinal health and immune status of suckling piglets. *Journal of Nutritional Science*. Cambridge University Press 4: e27-e27. DOI: 10.1017/jns.2015.16.

Heim, G., Sweeney, T., O'Shea, C. J., Doyle, D. N. and O'Doherty, J. V. (2015b). Effect of maternal dietary supplementation of laminarin and fucoidan, independently or in combination, on pig growth performance and aspects of intestinal health. *Animal Feed Science and Technology* 204: 28-41. DOI: 10.1016/j.anifeedsci.2015.02.007.

Herman, E. M. and Schmidt, M. A. (2016). The potential for engineering enhanced functional-feed soybeans for sustainable aquaculture feed. *Frontiers in Plant Science* 7: 440. Available at: https://www.frontiersin.org/article/10.3389/fpls.2016.00440.

Holdt, S. L. and Kraan, S. (2011). Bioactive compounds in seaweed: functional food applications and legislation. *Journal of Applied Phycology* 23(3): 543-597. DOI: 10.1007/s10811-010-9632-5.

Honya, M., Mori, H., Anzai, M., Araki, Y. and Nisizawa, K. (1999). Monthly changes in the content of fucans, their constituent sugars and sulphate in cultured *Laminaria japonica*. In: *Proceedings of the Sixteenth International Seaweed Symposium*, pp. 411-416. DOI: 10.1007/978-94-011-4449-0_49.

Jones, R. T., Blunden, G. and Probert, A. J. (1979). Effects of dietary Ascophyllum nodosum on blood parameters of rats and pigs. *Botanica Marina* 22: 393-402.

Jones, C. K., Gabler, N. K., Main, R. G. and Patience, J. F. (2012). Characterizing growth and carcass composition differences in pigs with varying weaning weights and postweaning performance1. *Journal of Animal Science* 90(11): 4072-4080. DOI: 10.2527/jas.2011-4793.

Khan, N., Ryu, K. Y., Choi, J. Y., Nho, E. Y., Habte, G., Choi, H., Kim, M. H., Park, K. S. and Kim, K. S. (2015). Determination of toxic heavy metals and speciation of arsenic in seaweeds from South Korea. *Food Chemistry* 169: 464-470. DOI: 10.1016/j.foodchem.2014.08.020.

Koreleski, J., Świątkiewicz, S. and Arczewska, A. (2010). The effect of dietary potassium and sodium on performance, carcass traits, and nitrogen balance and excreta moisture in broiler chicken. *Journal of Animal and Feed Sciences* 19(2): 244-256. DOI: 10.22358/jafs/66285/2010.

Kraan, S. (2013). 6 - Pigments and minor compounds in algae. In: Domínguez HBT-FI from A for F and N (Ed.) *Functional Ingredients from Algae for Foods and Nutraceuticals*. *Woodhead Publishing Series in Food Science, Technology and Nutrition*. Woodhead Publishing, pp. 205-251. DOI: 10.1533/9780857098689.1.205.

Lafarga, T., Acién-Fernández, F. G. and Garcia-Vaquero, M. (2020). Bioactive peptides and carbohydrates from seaweed for food applications: natural occurrence, isolation, purification, and identification. *Algal Research* 48. DOI: 10.1016/j.algal.2020.101909: 101909.

Lahaye, M. (1991). Marine algae as sources of fibres: determination of soluble and insoluble dietary fibre contents in some 'sea vegetables'. *Journal of the Science of Food and Agriculture*. John Wiley & Sons, Ltd 54(4): 587–594. DOI: 10.1002/jsfa.2740540410.

Landa-Cansigno, C., Hernández-Carmona, G., Arvizu-Higuera, D. L., Muñoz-Ochoa, M. and Hernández-Guerrero, C. J. (2017). Bimonthly variation in the chemical composition and biological activity of the brown seaweed Eisenia arborea (Laminariales: Ochrophyta) from Bahía Magdalena, Baja California Sur, Mexico. *Journal of Applied Phycology* 29(5): 2605–2615. DOI: 10.1007/s10811-017-1195-2.

Leonard, S. G., Sweeney, T., Bahar, B., Lynch, B. P. and O'Doherty, J. V. (2011). Effects of dietary seaweed extract supplementation in sows and post-weaned pigs on performance, intestinal morphology, intestinal microflora and immune status. *British Journal of Nutrition*. Cambridge University Press 106(5): 688–699. DOI: 10.1017/S0007114511000997.

Lourenço, S. O., Barbarino, E., De-Paula, J. C., Pereira, L. OdS. and Marquez, U. M. L. (2002). Amino acid composition, protein content and calculation of nitrogen-to-protein conversion factors for 19 tropical seaweeds. *Phycological Research*. John Wiley & Sons, Ltd 50(3): 233–241. DOI: 10.1111/j.1440-1835.2002.tb00156.x.

Lozano, I., Wacyk, J. M., Carrasco, J. and Cortez-San Martín, M. A. (2016). Red macroalgae Pyropia columbina and Gracilaria chilensis: sustainable feed additive in the Salmo salar diet and the evaluation of potential antiviral activity against infectious salmon anemia virus. *Journal of Applied Phycology* 28(2): 1343–1351. DOI: 10.1007/s10811-015-0648-8.

Mæhre, H. K., Malde, M. K., Eilertsen, K. E. and Elvevoll, E. O. (2014). Characterization of protein, lipid and mineral contents in common Norwegian seaweeds and evaluation of their potential as food and feed. *Journal of the Science of Food and Agriculture*. John Wiley & Sons, Ltd 94(15): 3281–3290. DOI: 10.1002/jsfa.6681.

Makkar, H. P. S., Tran, G., Heuzé, V., Giger-Reverdin, S., Lessire, M., Lebas, F. and Ankers, P. (2016). Seaweeds for livestock diets: a review. *Animal Feed Science and Technology* 212: 1–17. DOI: 10.1016/j.anifeedsci.2015.09.018.

Manns, D., Nielsen, M. M., Bruhn, A., Saake, B. and Meyer, A. S. (2017). Compositional variations of brown seaweeds Laminaria digitata and Saccharina latissima in Danish waters. *Journal of Applied Phycology* 29(3): 1493–1506. DOI: 10.1007/s10811-017-1056-z.

McAlpine, P., O'Shea, C. J., Varley, P. F., Flynn, B. and O'Doherty, J. V. (2012). The effect of seaweed extract as an alternative to zinc oxide diets on growth performance, nutrient digestibility, and fecal score of weaned piglets. *Journal of Animal Science* 90 (Suppl. 4): 224–226. DOI: 10.2527/jas.53956.

McDonnell, P., Figat, S. and O'Doherty, J. V. (2010). The effect of dietary laminarin and fucoidan in the diet of the weanling piglet on performance, selected faecal microbial populations and volatile fatty acid concentrations. *Animal* Cambridge University Press 4(4): 579–585. DOI: 10.1017/S1751731109991376.

Michalak, I. and Chojnacka, K. (2014). Algal extracts: technology and advances. *Engineering in Life Sciences*. John Wiley & Sons, Ltd 14(6): 581–591. DOI: 10.1002/elsc.201400139.

Michiels, J., Skrivanova, E., Missotten, J., Ovyn, A., Mrazek, J., De Smet, S. and Dierick, N. (2012). Intact brown seaweed (Ascophyllum nodosum) in diets of weaned piglets: effects on performance, gut bacteria and morphology and plasma oxidative status. *Journal of Animal Physiology and Animal Nutrition*. John Wiley & Sons, Ltd 96(6): 1101–1111. DOI: 10.1111/j.1439-0396.2011.01227.x.

Miranda, M., Lopez-Alonso, M. and Garcia-Vaquero, M. (2017). Macroalgae for functional feed development: applications in aquaculture, ruminant and swine feed industries. In: *Seaweeds: Biodiversity, Environmental Chemistry and Ecological Impacts*, pp. 133–153. Available at: https://www.scopus.com/inward/record.uri?eid=2-s2.0-850 48822636&partnerID=40&md5=97e44f36e4ee7081f3c8bb790bae7eba.

Mols-Mortensen, A., Ortind, E. áG., Jacobsen, C. and Holdt, S. L. (2017). Variation in growth, yield and protein concentration in Saccharina latissima (Laminariales, Phaeophyceae) cultivated with different wave and current exposures in the Faroe Islands. *Journal of Applied Phycology* 29(5): 2277–2286. DOI: 10.1007/s10811-017-1169-4.

Morais, T., Inácio, A., Coutinho, T., Ministro, M., Cotas, J., Pereira, L. and Bahcevandziev, K. (2020). Seaweed potential in the animal feed: a review. *Journal of Marine Science and Engineering*. MDPI AG 8(8). DOI: 10.3390/JMSE8080559.

Moroney, N. C., O'Grady, M. N., O'Doherty, J. V. and Kerry, J. P. (2012). Addition of seaweed (Laminaria digitata) extracts containing laminarin and fucoidan to porcine diets: influence on the quality and shelf-life of fresh pork. *Meat Science* 92(4): 423–429. DOI: 10.1016/j.meatsci.2012.05.005.

Moroney, N. C., O'Grady, M. N., Robertson, R. C., Stanton, C., O'Doherty, J. V. and Kerry, J. P. (2015). Influence of level and duration of feeding polysaccharide (laminarin and fucoidan) extracts from brown seaweed (Laminaria digitata) on quality indices of fresh pork. *Meat Science* 99: 132–141. DOI: 10.1016/j.meatsci.2014.08.016.

Nielsen, M. M., Manns, D., D'Este, M., Krause-Jensen, D., Rasmussen, M. B., Larsen, M. M., Alvarado-Morales, M., Angelidaki, I. and Bruhn, A. (2016). Variation in biochemical composition of Saccharina latissima and Laminaria digitata along an estuarine salinity gradient in inner Danish waters. *Algal Research* 13: 235–245. DOI: 10.1016/j.algal.2015.12.003.

O'Doherty, J. V., McDonnell, P. and Figat, S. (2010). The effect of dietary laminarin and fucoidan in the diet of the weanling piglet on performance and selected faecal microbial populations. *Livestock Science* 134(1-3): 208–210. DOI: 10.1016/j.livsci.2010.06.143.

O'Shea, C. J., McAlpine, P., Sweeney, T., Varley, P. F. and O'Doherty, J. V. (2014). Effect of the interaction of seaweed extracts containing laminarin and fucoidan with zinc oxide on the growth performance, digestibility and faecal characteristics of growing piglets. *British Journal of Nutrition*. Cambridge University Press 111(5): 798–807. DOI: 10.1017/S0007114513003280.

Øverland, M., Mydland, L. T. and Skrede, A. (2019). Marine macroalgae as sources of protein and bioactive compounds in feed for monogastric animals. *Journal of the Science of Food and Agriculture*. John Wiley & Sons, Ltd 99(1): 13–24. DOI: 10.1002/jsfa.9143.

Rajauria, G. (2018). Optimization and validation of reverse phase HPLC method for qualitative and quantitative assessment of polyphenols in seaweed. *Journal* of *Pharmaceutical and Biomedical Analysis* 148: 230-237. DOI: 10.1016/j.jpba.2017.10.002.

Rajauria, G., Draper, J., McDonnell, M. and O'Doherty, J. V. (2016). Effect of dietary seaweed extracts, galactooligosaccharide and vitamin E supplementation on meat quality parameters in finisher pigs. *Innovative Food Science and Emerging Technologies* 37: 269-275. DOI: 10.1016/j.ifset.2016.09.007.

Rattigan, R., Sweeney, T., Vigors, S., Thornton, K., Rajauria, G. and O'Doherty, A. J. V. (2019). The effect of increasing inclusion levels of a fucoidan-rich extract derived from Ascophyllum nodosum on growth performance and aspects of intestinal health of pigs post-weaning. *Marine Drugs* 17(12) . DOI: 10.3390/md17120680.

Rattigan, R., Sweeney, T., Maher, S., Thornton, K., Rajauria, G. and O'Doherty, J. V. (2020). Laminarin-rich extract improves growth performance, small intestinal morphology, gene expression of nutrient transporters and the large intestinal microbial composition of piglets during the critical post-weaning period. *British Journal of Nutrition*. Cambridge University Press 123(3): 255-263. DOI: 10.1017/S0007114519002678.

Regulation EC (2011) Commission Regulation (EU) No 574/2011 of 16 June 2011 amending Annex I to Directive 2002/32/EC of the European Parliament and of the Council as regards maximum levels for nitrite, melamine. Ambrosia spp. and carry-over of certain coccidiostats and histom. *Official Journal of the European Union, L* 159: 7-24.

Rey-Crespo, F., López-Alonso, M. and Miranda, M. (2014). The use of seaweed from the Galician coast as a mineral supplement in organic dairy cattle. *Animal* 8(4): 580-586. DOI: 10.1017/S1751731113002474.

Rioux, L. E., Turgeon, S. L. and Beaulieu, M. (2010). Structural characterization of laminaran and galactofucan extracted from the brown seaweed Saccharina longicruris. *Phytochemistry* 71(13): 1586-1595. DOI: 10.1016/j.phytochem.2010.05.021.

Roleda, M. Y., Marfaing, H., Desnica, N., Jónsdóttir, R., Skjermo, J., Rebours, C. and Nitschke, U. (2019). Variations in polyphenol and heavy metal contents of wild-harvested and cultivated seaweed bulk biomass: health risk assessment and implication for food applications. *Food Control* 95: 121-134. DOI: 10.1016/j.foodcont.2018.07.031.

Ronan, J. M., Stengel, D. B., Raab, A., Feldmann, J., O'Hea, L., Bralatei, E. and McGovern, E. (2017). High proportions of inorganic arsenic in Laminaria digitata but not in Ascophyllum nodosum samples from Ireland. *Chemosphere* 186: 17-23. DOI: 10.1016/j.chemosphere.2017.07.076.

Rose, M., Lewis, J., Langford, N., Baxter, M., Origgi, S., Barber, M., MacBain, H. and Thomas, K. (2007). Arsenic in seaweed—forms, concentration and dietary exposure. *Food and Chemical Toxicology* . Elsevier 45(7): 1263-1267.

Ruiz, Á. R., Gadicke, P., Andrades, S. M. and Cubillos, R. (2018). Supplementing nursery pig feed with seaweed extracts increases final body weight of pigs. *Austral Journal of Veterinary Sciences*. Universidad Austral de Chile. Facultad de Ciencias Veterinarias 50(2): 83-87. DOI: 10.4067/S0719-81322018000200083.

Sauvageau, C. (Camille) (1920). *Utilisation Des Algues Marines*. Paris: O. Doin. Available at: https://www.biodiversitylibrary.org/item/122691.

Schiener, P., Black, K. D., Stanley, M. S. and Green, D. H. (2015). The seasonal variation in the chemical composition of the kelp species Laminaria digitata, Laminaria Hyperborea, Saccharina latissima and Alaria esculenta. *Journal of Applied Phycology* 27(1): 363-373. DOI: 10.1007/s10811-014-0327-1.

Schiener, P., Zhao, S., Theodoridou, K., Carey, M., Mooney-McAuley, K. and Greenwell, C. (2017). The nutritional aspects of biorefined Saccharina latissima, Ascophyllum nodosum and Palmaria palmata. *Biomass Conversion and Biorefinery* 7(2): 221-235. DOI: 10.1007/s13399-016-0227-5.

Silva, D. M., Valente, L. M. P., Sousa-Pinto, I., Pereira, R., Pires, M. A., Seixas, F. and Rema, P. (2015). Evaluation of IMTA-produced seaweeds (Gracilaria, Porphyra, and Ulva) as dietary ingredients in Nile tilapia, Oreochromis niloticus L., juveniles. Effects on growth performance and gut histology. *Journal of Applied Phycology* 27(4): 1671-1680. DOI: 10.1007/s10811-014-0453-9.

Stévant, P., Marfaing, H., Rustad, T., Sandbakken, I., Fleurence, J. and Chapman, A. (2017). Nutritional value of the kelps Alaria esculenta and Saccharina latissima and effects of short-term storage on biomass quality. *Journal of Applied Phycology* 29(5): 2417-2426. DOI: 10.1007/s10811-017-1126-2.

Stiger-Pouvreau, V., Bourgougnon, N. and Deslandes, E. (2016). Carbohydrates from seaweeds. In: Fleurence, J. (Translator) and Levine IBT-S in H and DP (eds) *Seaweed in Health and Disease Prevention*. San Diego: Academic Press, pp. 223-274. DOI: 10.1016/B978-0-12-802772-1.00008-7.

Suetsuna, K. and Nakano, T. (2000). Identification of an antihypertensive peptide from peptic digest of wakame (Undaria pinnatifida). *The Journal of Nutritional Biochemistry* 11(9): 450-454. DOI: 10.1016/S0955-2863(00)00110-8.

Swanson, J. C. (1995). Farm animal well-being and intensive production systems2. *Journal of Animal Science* 73(9): 2744-2751. DOI: 10.2527/1995.7392744x.

Sweeney, T. and O'Doherty, J. V. (2016). Marine macroalgal extracts to maintain gut homeostasis in the weaning piglet. *Domestic Animal Endocrinology* 56 (Suppl.): S84-S89. DOI: 10.1016/j.domaniend.2016.02.002.

Van Boeckel, T. P., Brower, C., Gilbert, M., Grenfell, B. T., Levin, S. A., Robinson, T. P., Teillant, A. and Laxminarayan, R. (2015). Global trends in antimicrobial use in food animals. *Proceedings of the National Academy of Sciences of the United States of America*. National Academy of Sciences 112(18): 5649-5654. DOI: 10.1073/pnas.1503141112.

Vigors, S., O'Doherty, J. V., Rattigan, R., McDonnell, M. J., Rajauria, G. and Sweeney, T. (2020). Effect of a laminarin rich macroalgal extract on the caecal and colonic microbiota in the post-weaned pig. *Marine Drugs* 18(3). DOI: 10.3390/md18030157.

Vigors, S., O'Doherty, J., Rattigan, R. and Sweeney, T. (2021). Effect of supplementing seaweed extracts to pigs until d35 post-weaning on performance and aspects of intestinal health. *Marine Drugs* 19(4) . DOI: 10.3390/md19040183.

Walsh, A. M., Sweeney, T., O'Shea, C. J., Doyle, D. N. and O'Doherty, J. V. (2013a). Effect of dietary laminarin and fucoidan on selected microbiota, intestinal morphology and immune status of the newly weaned pig. *British Journal of Nutrition*. Cambridge University Press 110(9): 1630-1638. DOI: 10.1017/S0007114513000834.

Walsh, A. M., Sweeney, T., O'Shea, C. J., Doyle, D. N. and O'Doherty, J. V. O. (2013b). Effect of supplementing varying inclusion levels of laminarin and fucoidan on growth performance, digestibility of diet components, selected faecal microbial populations and volatile fatty acid concentrations in weaned pigs. *Animal Feed Science and Technology* 183(3-4): 151-159. DOI: 10.1016/j.anifeedsci.2013.04.013.

Wan, J., Jiang, F., Xu, Q., Chen, D. and He, J. (2016). Alginic acid oligosaccharide accelerates weaned pig growth through regulating antioxidant capacity, immunity and intestinal development. *RSC Advances*. The Royal Society of Chemistry 6(90): 87026–87035. DOI: 10.1039/C6RA18135J.

Wan, J., Zhang, J., Chen, D., Yu, B. and He, J. (2017). Effects of alginate oligosaccharide on the growth performance, antioxidant capacity and intestinal digestion-absorption function in weaned pigs. *Animal Feed Science and Technology* 234: 118–127. DOI: 10.1016/j.anifeedsci.2017.09.006.

Wan, J., Zhang, J., Chen, D., Yu, B., Huang, Z., Mao, X., Zheng, P., Yu, J. and He, J. (2018). Alginate oligosaccharide enhances intestinal integrity of weaned pigs through altering intestinal inflammatory responses and antioxidant status. *RSC Advances*. The Royal Society of Chemistry 8(24): 13482–13492. DOI: 10.1039/C8RA01943F.

Wang, T., Jónsdóttir, R., Kristinsson, H. G., Hreggvidsson, G. O., Jónsson, J. Ó, Thorkelsson, G. and Ólafsdóttir, G. (2010). Enzyme-enhanced extraction of antioxidant ingredients from red algae Palmaria palmata. *LWT—Food Science and Technology* 43(9): 1387–1393. DOI: 10.1016/j.lwt.2010.05.010.

Wilkinson, M. (1992). *Seaweed Resources in Europe: Uses and Potential*. Guiry, M. D. and Blunden, G. (Eds). Chichester: John Wiley & Sons Ltd, 1991. xi+432pp. Price: £65.00. ISBN 0 471 92947 6. Aquatic Conservation: Marine and Freshwater Ecosystems 2(2). John Wiley & Sons, Ltd: 209–210. DOI: 10.1002/aqc.3270020206.

Yamada, M., Yamamoto, K., Ushihara, Y. and Kawai, H. (2007). Variation in metal concentrations in the brown alga Undaria pinnatifida in Osaka Bay, Japan. *Phycological Research*. John Wiley & Sons, Ltd 55(3): 222–230. DOI: 10.1111/j.1440-1835.2007.00465.x.

Yuan, Y., Zhang, J., Fan, J., Clark, J., Shen, P., Li, Y. and Zhang, C. (2018). Microwave assisted extraction of phenolic compounds from four economic brown macroalgae species and evaluation of their antioxidant activities and inhibitory effects on α-amylase, α-glucosidase, pancreatic lipase and tyrosinase. *Food Research International* 113: 288–297. DOI: 10.1016/j.foodres.2018.07.021.

www.ingramcontent.com/pod-product-compliance
Lightning Source LLC
Chambersburg PA
CBHW050528270326
41926CB00015B/3120